Human Biology
Lab Manual

Werner Williams

St. John's River State College

Cover images © Shutterstock.com

www.kendallhunt.com
Send all inquiries to:
4050 Westmark Drive
Dubuque, IA 52004-1840

Copyright © 2017, 2018 by Kendall Hunt Publishing Company

ISBN 978-1-5249-6153-4

All rights reserved. No part of this publication may be reproduced, stored in a retrieval system, or transmitted, in any form or by any means, electronic, mechanical, photocopying, recording, or otherwise, without the prior written permission of the copyright owner.

Published in the United States of America.

Contents

Lab 1: The Use and Care of the Microscope

Many biological objects are so small that a microscope is needed to see them. Light microscopes use light rays that are focused using magnifying lenses.

How a Microscope Works

Light is released from a light source and is directed by a condenser lens onto the specimen which is usually found on a slide. The light from the specimen then passes into an objective lens, which magnifies and bends the light rays and sends it to a projector lens, which reverses the direction of the rays so that when it reaches the eye it will not appear to be upside down. As the light rays travel to the eyepiece the image is further magnified and projected into the eye. (Figure 1)

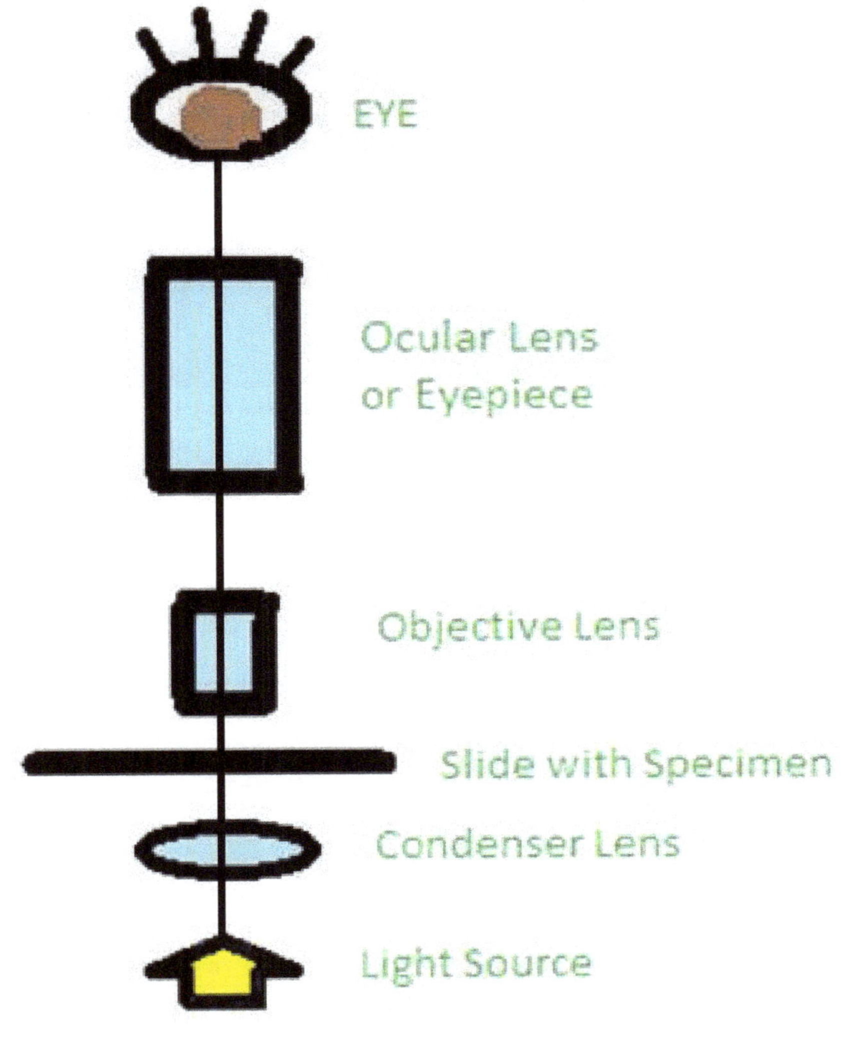

Source: Werner Williams

Figure 1

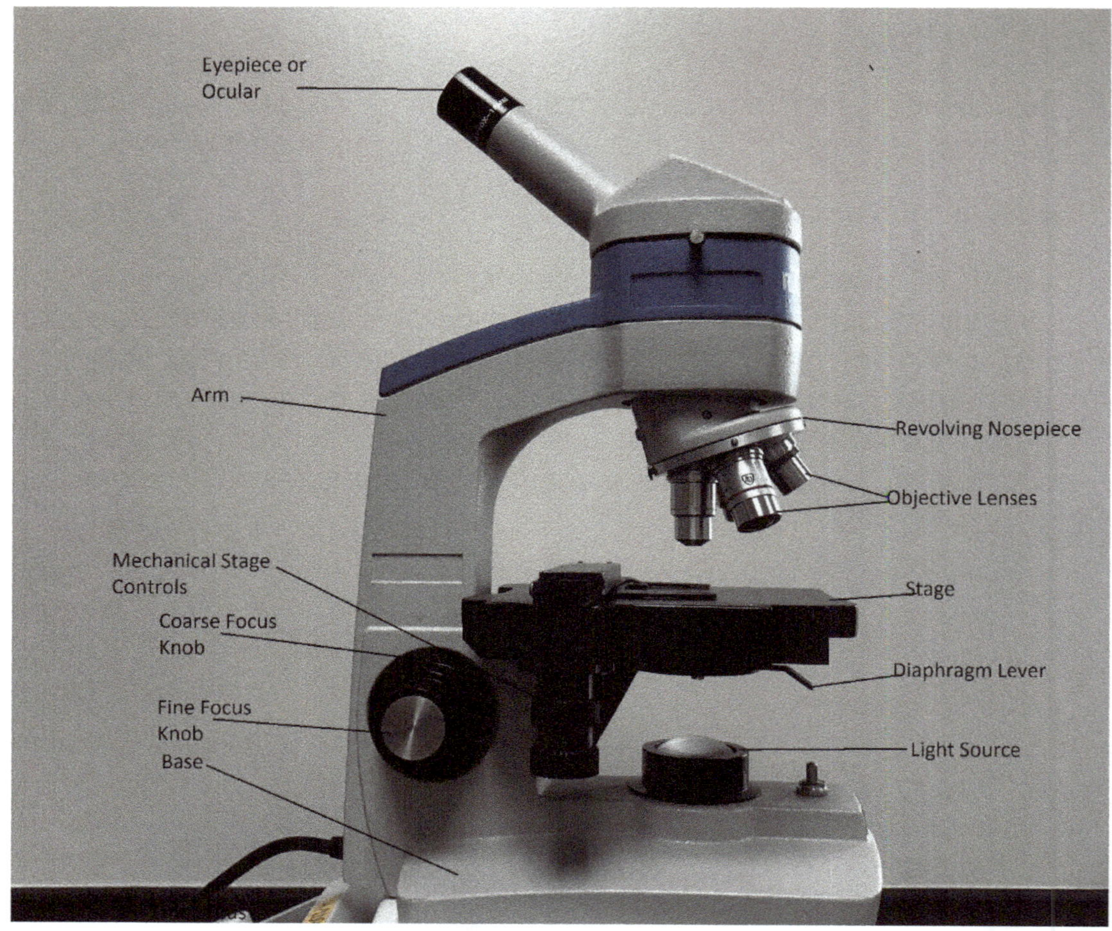

Source: Werner Williams

Figure 2 A Typical Light Microscope

Table 1 Microscope Parts and Their Functions

Parts	Function
Eyepiece or Ocular	Topmost series of lenses through which we look to view the specimen.
Arm	This connects the upper parts to the base.
Switch	This turns the light on or off.
Coarse Focus Knob	This large knob is used to bring the specimen into approximate focus. It is only used with the scanning and low power objective lenses.
Fine Focus Knob	This is used to fine tune the focus on the specimen.
Base	The flat part of the microscope that rests on the table.
Mechanical Stage Control Knobs	The two knobs located beneath the stage that control the movement of the slide. One controls the forward/reverse movement and the other controls the left/right movement.

Parts	Function
Stage	The flat platform on which the slide is placed.
Objective Lenses	The lenses closest to the specimen with the magnification values (4x, 10x, 43x, and 100x) engraved on the objectives. *Scanning*-Lens with the least magnification (usually 4x). *Low Power*- Lens with a greater magnification than the scanning objective (usually 10x). *High Power*- Lens with a greater magnification than the low power objective (usually 40x). *Oil Immersion*- Lens with the greatest magnification (usually 100x).
Revolving Nosepiece	The part of the microscope that holds the objective lenses.
Diaphragm Control Lever	This lever varies the amount of light passing through the stage opening. This helps illuminate the specimen and helps in contrast and resolution.
Light Source	A lamp that sends a beam of light through the specimen

Source: Werner Williams

Magnification

Compound microscopes contain at least two lenses. The first lens would be the **eyepiece,** also called the **ocular**, and usually magnifies objects 10 times (10X). The other lenses are the **objective lens**, which vary in their magnification. They usually range from 4X to 100X. The 100X lens usually called the oil immersion lens is used for viewing very small objects such as bacteria. Prior to using this lens, a drop of immersion oil is placed on the slide.

Contrast

Contrast refers to the darkness of the background relative to the specimen. There must be sufficient contrast between the different parts of the object you are viewing for you to distinguish those different parts. Sometimes stains are added to the specimen to increase the contrast. Reducing the amount of light improves the contrast when viewing unstained specimens.

Resolving Power

A good microscope not only magnifies objects but also allows us to **resolve** or distinguish objects. A good light microscope can resolve objects that are about 0.5μm or more apart.

Focusing

As a general rule, you should start focusing with the scanning (4X) or low power lens (10X) and then use the high power objective lens (40X). Starting with a lower power lens allows you to more easily view and magnify a larger part of the specimen. When a larger power lens is used, you are actually seeing a smaller part of the specimen.

Working Distance

This is the space between the lens of the microscope and the top of the specimen. This is the free area available to work and manipulate the specimen. Of the three objectives shown in Figure 3, the 10X lens has the largest working distance and the 100X lens has the smallest.

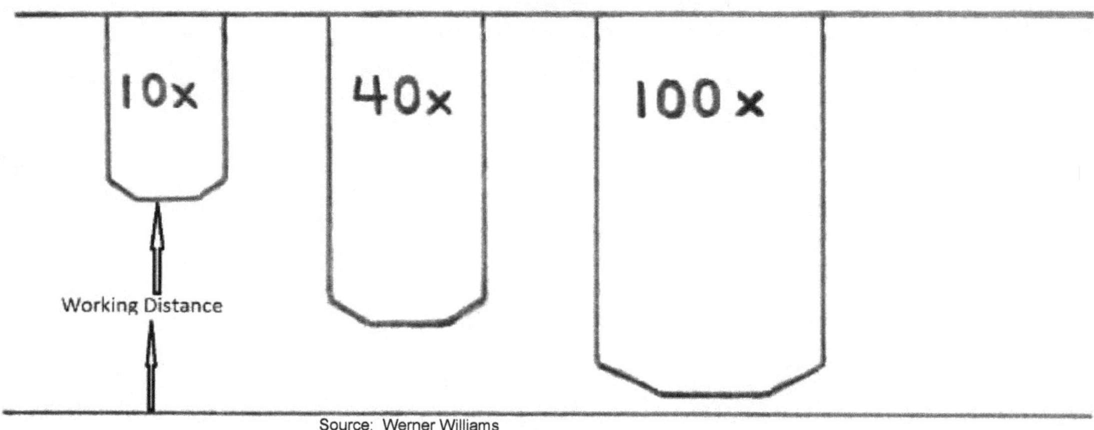

Figure 3 Working Distance

Parfocal

Your microscope is parfocal, meaning that if an image is in focus under low power and you wish to switch to a higher power, when you switch, it should still roughly be in focus. You should only have to make small adjustments with the fine adjustment knob to see a clear image.

Rules for Microscope Use

1. Use both hands when carrying the microscope.
2. The lowest power objective should be in position both at the beginning and end of microscope use.
3. Use only lens paper for cleaning the lenses.
4. Do not remove parts of the microscope.
5. Report any malfunctioning.

Total Magnification

The total magnification is calculated by multiplying the power of the ocular (5X or 10X) by the power of the objective lens.

Total Magnification = Ocular lens power x Objective lens power

Magnification of your ocular lens alone=

Complete the following table showing the total magnification using each objective.

Table 2 Total Magnification

	Magnification of Each Objective Lens Alone	Total Magnification (Objective x Ocular)
Scanning		
Low Power		
High Power		
Oil Immersion		

Source: Werner Williams

Diameter of the Field of View

The **field** is the lighted circle you see when you look through the eyepiece. You will use a ruler to help estimate the field of view.

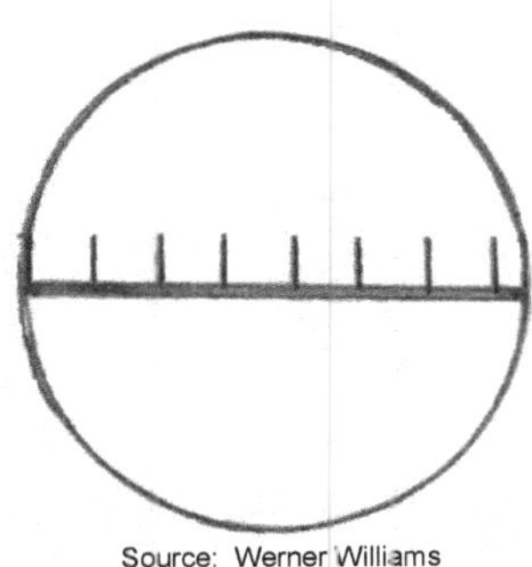

Source: Werner Williams

Figure 4 View of Ruler Through a Microscope

Procedure 1 Diameter of the Field of View

1. Place a clear plastic ruler on top of the metal clips that are used to hold a slide on the stage.
2. Place the millimeter scale over the center of the stage opening.
3. Click the 10X objective lens into place.
4. Use the mechanical stage control knobs to move the ruler so as to locate the millimeter lines of the ruler and place these lines in the middle of the field of view.
5. Use the coarse focus knob to bring the lines into focus. The distance between two lines on the ruler represents 1mm.
6. While looking through the eyepiece, set the mechanical stage control knobs to move the ruler so that one of the millimeter lines is just touching the left side of the field of view. Figure 4.
7. Count the number of millimeter lines (actually spaces) that are visible. Estimate to the nearest tenth of a millimeter. What is the diameter?_________
   Convert that into micrometers (μm). Use the formula below:
   1mm=1000 μm; So 1.3 mm=1,300 μm
   Low power lens (10X, 100X total) field of view:______mm or__________ μm

Because under high power (40x, 400x total) the thickness of one of the millimeter lines takes up practically the entire field of view, it is difficult to estimate the diameter of the field view under high power magnification.

The diameter under high power can be calculated on paper. Record the following data for your microscope:

1. LPD=low-power diameter of field (in micrometers)____________
2. LPM=low-power total magnification._________________

3. HPM=high-power total magnification_______________
4. Compute the high-power diameter of field (HPD) by plugging in the data into the following formula:

$$HPD = LPD \times (LPM / HPM)$$

High-power objective (40x, 400 total field of view):____________μm

Does low power or high power have a larger field of view and allow you to see more of the object?_______________________

Which has a smaller field but magnifies to a greater extent, 10X, 40X or 100X?

Depth of Field

When we view objects with a microscope, we are viewing those objects from above. The distance between the closest and farthest objects in focus within a scene is called the depth of field. You will be using a slide with three colored threads similar to what you see in Figure 5.

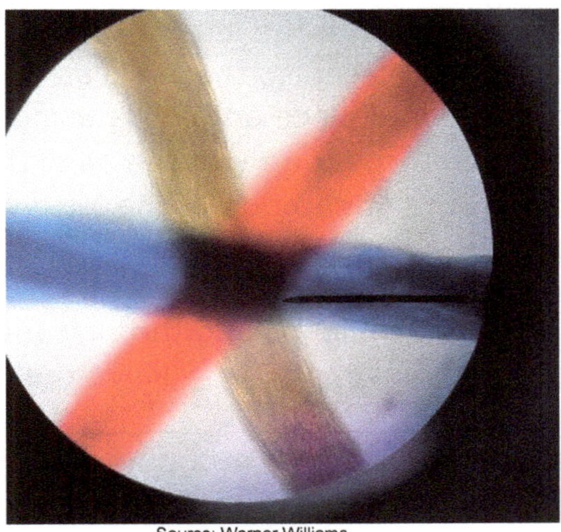

Source: Werner Williams

Figure 5 Three Colored Threads

Procedure 2 Depth of Field

1. Obtain a slide with three crossed colored threads and focus using the low power objective and the coarse focus knob.
2. Move the slide so that the area where all three threads overlap is in the center of your field of view and focus.
3. Switch to the high power lens and use the fine focus knob to fine-tune the image. *Never use the coarse focus knob with the high-power or oil immersion objectives.* Move the slide so you can see all three colors at once.
4. Looking through the microscope, turn the fine focus knob toward you to lower the stage just until the threads are out of focus. Now, slowly roll the stage upward noting which color thread comes into clear focus first, then second, then last. The first thread to come into focus when you raise the stage is the one that is on top. The second thread to come into focus is the one in the middle.
5. Record your observations, relative to which color of thread is on the top, middle, and bottom.
   Top thread color ______________________________
   Middle thread color ____________________________
   Bottom thread color ____________________________

Remember, *never start observations with the high power or oil immersion objectives. Instead, start at a lower power and work up to the 40X or 100X objectives.*

Procedure 3 Viewing the Letter "e"

1. Obtain a microscope slide with the letter "e."
2. Place the slide on the stage with the letter over the circular opening in the stage. The slide should be placed so that "e" can be read with the naked eye.
3. Rotate the 10X objective into place and focus using the coarse focus knob.
4. Center the letter "e" and focus. Can you see all of the "e"? ___________ What is different about the orientation of the letters when viewed with the microscope instead of the naked eye?

   ____
5. Move the slide to the right while looking through the eyepiece. In which direction does the image move?_________Move the image to the left. In which direction does it move? ___________________________
6. Center the "e", focus and rotate the 40X objective into place and focus using the fine focus knob. Remember *never use the coarse focus knob with the high power or oil immersion objectives*. You should be seeing "ink blotches" that compose the "e".
7. Draw the letter "e" when observed with each of the following:

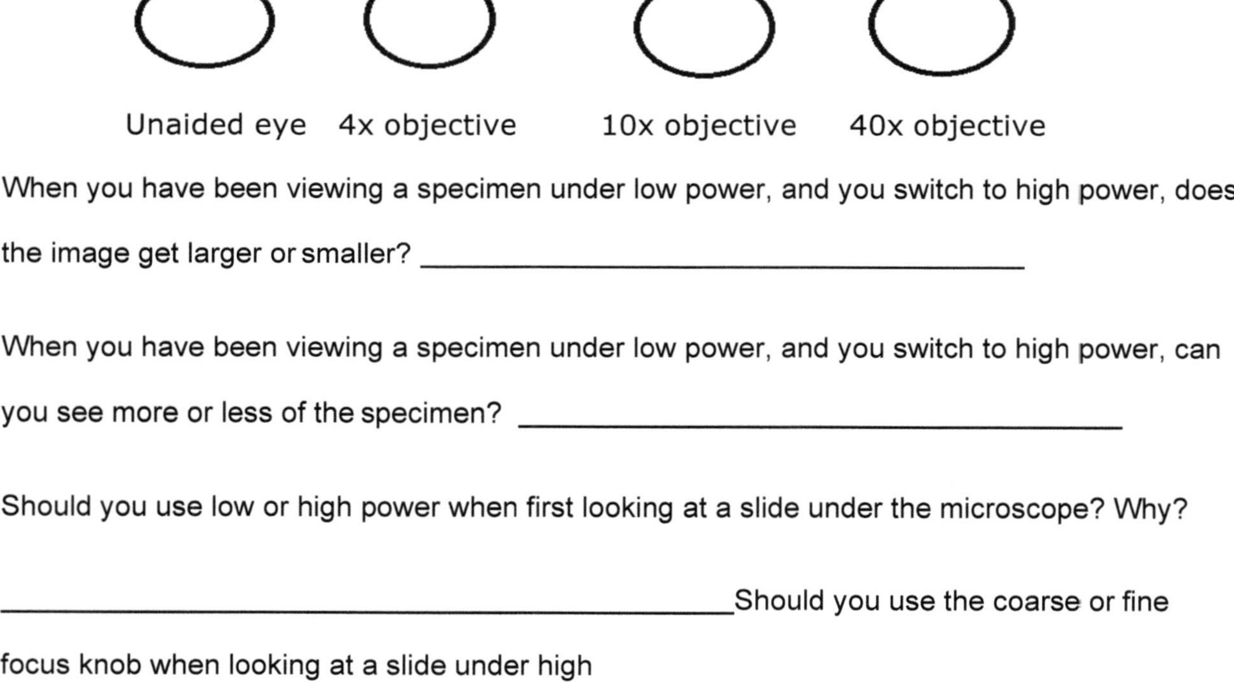

When you have been viewing a specimen under low power, and you switch to high power, does the image get larger or smaller? ___________________________________

When you have been viewing a specimen under low power, and you switch to high power, can you see more or less of the specimen? ___________________________________

Should you use low or high power when first looking at a slide under the microscope? Why?

____Should you use the coarse or fine focus knob when looking at a slide under high power?____

Preparing a Wet Mount

In a wet mount, a specimen is placed in a drop of water or other liquid and then covered with a coverslip. Figure 6. You will be performing two wet mounts; one of your cheek cells and the other of a slice of onion.

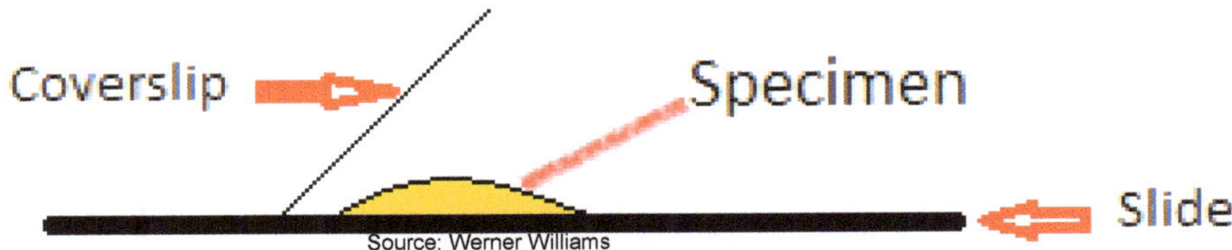

Figure 6 Wet Mount

Procedure 4 Wet Mount of Cheek Cells

1. Obtain a clean microscope slide and a cover slip.
2. Place a small drop of iodine on the slide.
3. Obtain a clean toothpick from your instructor and gently scrape the inside of your cheek about ten times.
4. Roll the toothpick with the cheek scrapings into the drop of stain on your slide and stir the contents gently.
5. Discard the tooth-pick.
6. Carefully lower a cover slip over the stain/cheek scrapings mixture. Note: slowly lowering the cover slip at about a 45°degree angle helps avoid air bubbles getting trapped beneath the cover slip.
7. Observe your cheek scrapings under the low power lens (10X), and the high power lens (40X). Remember to put the cells in the center of your field of view before increasing the magnification.
8. Find a few isolated cheek cells and focus on them. (Ignore clumps of cells). You should see roundish cells, with a stained nucleus in the center.
9. Draw a couple of your cells seen under high power below and then discard the slide and coverslip in the beaker of bleach.

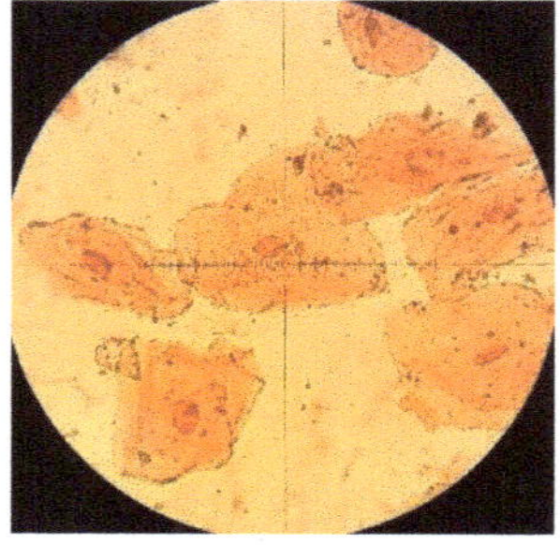

© Thawatchai Tumwapee/Shutterstock.com

Figure 7 Cheek Cells

Procedure 5 Wet Mount of Onion Cells.

1. Obtain a clean microscope slide and coverslip.
2. With a scalpel or your fingers, strip a thin, transparent layer of cells from a piece of onion.
3. Place it flat on the slide.
4. Add a drop of iodine and cover with a coverslip.
5. Observe the cells under the low power lens (10X), and the high power lens (40X). Be sure to label your drawing.
6. Locate the cell wall. Is a nucleus visible? One to three nucleoli may be visible within the nucleus.
7. Draw a couple of the cells on high power and label the cell wall, nucleus and cytoplasm.
8. Record some obvious differences between the human cheek cells and the onion cells. ____

____

9. Record any similarities between the two types of cells. ________________

____

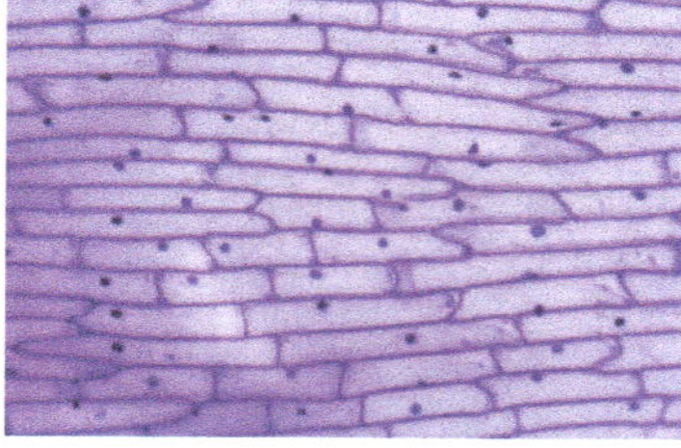

© Wave_Movies/Shutterstock.com

Figure 8 Onion cells

10. Discard the slide and coverslip in the beaker of bleach.

Lab 2: The Scientific Method

Science can be used to answer many common everyday questions such as Why is my flashlight not working? Why do some cars get better gas mileage than others? Can my diet cause heart disease? Can a particular exercise keep my weight off?

To answer questions, Scientists have developed a series of steps called the **scientific method**. The steps are as follows:

1) **Observation.** Collect all pertinent information.
2) **Question.** Ask a question about what you have observed.
3) **Hypothesis.** Suggest a possible answer or hunch called a hypothesis. This must be testable & falsifiable (can be proven false).
4) **Prediction.** To determine if the hypothesis is correct, a prediction should be made that can be tested. This is called "if-then" reasoning.
5) **Experiment.** Conduct an experiment to see if the prediction is correct.
6) **Conclusion**. Evaluate the results to determine if the hypothesis is correct.

Experiments should be so well designed and described, that others should be able to conduct the experiment and get the same results.

In some experiments, the subjects or individuals that are being tested may be divided into two groups: an **experimental group** that is treated with the **experimental variable** and a **control group** that is not.

There are three types of variables: the **independent variable** is the one factor that we are testing. The **dependent variable** is the result that is measured or observed at the end of the experiment. The **controlled variables** are all the other factors which we keep the same for all the groups being studied.

To investigate the question, "do plants grow taller with fertilizer", our hypothesis may be, "Plants grown with fertilizer grow taller than plants that don't get fertilizer". The prediction may be that if we expose plants to fertilizer, they will grow taller than plants not exposed to fertilizer.

So we may obtain 20 plants. They should be as genetically identical as possible and about the same height. We could divide the plants into 2 groups: 10 plants are the control group and the other 10 are the experimental group.

All twenty plants are treated the same. For example, they get the same amount of water, the same air, the same light. These factors are the **controlled variables**. Then we introduce the **independent variable**. That is the one factor we are testing, which in this experiment would be the effect of the fertilizer. The experimental group receives fertilizer while the control group does not.

The experiment is then conducted for a period of time and then the height of the plants are measured. Since height is the result we are measuring, it is the **dependent variable**. If the results show that the experimental plants grew taller, then the hypothesis is proven correct. If both sets of plants grew to the same height, then the hypothesis is incorrect. Discovering that a hypothesis is incorrect is just as valuable as showing that it is correct.

A well designed experiment should be able to prove or disprove the hypothesis.

Table 1 has three observations that may be testable.

Table 1

Observation	**Question**	**Hypothesis**	**Prediction**
1. Heart rate is steady when we are at rest sitting.	If we are standing or holding our breath, will our heart rate speed up?	Our heart rate will stay the same whether we are sitting, standing or holding our breath.	If the hypothesis is correct, our heart rate should stay the same, whether we are sitting, standing or holding our breath.
2. We can hold our breath for a short period of time.	Can we hold our breath longer after we exercise?	We can hold our breath longer immediately after we exercise.	If the hypothesis is correct, after we exercise, we should be able to hold our breath for a longer period of time.
3. Tall people have long arms.	Do tall people have longer arms than shorter people?	The taller the person, the longer their arm length.	If the hypothesis is correct, taller people should have longer arm lengths than shorter people.

Source: Werner Williams

Choose along with the members in your group one prediction/hypothesis from Table 1 to conduct as an experiment. Consider the next three paragraphs below and the procedures you would use before choosing which one to test.

To test the first prediction, the test subjects should have their pulse checked at their wrist (which reveals the heart rate) while sitting, sitting and holding their breath, and then while standing.

To test the second prediction, the test subjects will see how long they can hold their breath at rest while sitting and then after exercising.

To test the third prediction, the test subject's height and arm length should be measured and compared.

Procedures

Hypothesis #1 (Pulse Rate)

1. Invite at least 8 students to be your test subjects. They should be as varied as possible. Include equal numbers of males and females. Include yourselves.
2. Have the subjects sit quietly, find the pulse on their wrist, and using a timer count the number of beats for 1 minute. **Do not take your own pulse.**
3. Repeat step 2 for each subject while the person stands.
4. Repeat step 2 for each subject but have the person sit while holding their breath. (If you can't hold your breath for 1 minute, hold it for 30 seconds and multiply by 2).
5. Record the results.

Hypothesis #2 (Holding Breath)

1. Invite at least 8 students to be your test subjects. They should be as varied as possible. Include equal numbers of males and females. Include yourselves.
2. Obtain a timer and while the subjects are sitting, have them hold their breath as long as they can while timing them.
3. Have the subjects walk down and up the stairs **twice** very quickly, and immediately on entering the lab, have them sit, hold their breath while timing them and record the results.

Hypothesis #3 (Height and Arm Length)

1) Invite at least 8 students to be your test subjects. They should be as varied as possible. Include equal numbers of males and females. Include yourselves.
2) Have your subjects stand and using a meter stick, measure their height in cm.
3) Using a tape measure, measure the length of both arms in cm, add the lengths and divide by two to get that person's average arm length.
4) Record the results.

Hypothesis to be tested____

____

Notify your instructor of the experiment you will be conducting before you begin.

RESULTS

Table 2 Pulse Rate Results

Test Subject	Heart Rate Sitting	Heart Rate Standing	Heart Rate While Holding Breath
1			
2			
3			
4			
5			
6			
7			
8			

Source: Werner Williams

Table 3 Holding Breath Results

Test Subjects	Holding Breath at Rest	Holding Breath after Exercise
1		
2		
3		
4		
5		
6		
7		
8		

Source: Werner Williams

Table 4 Height and Arm Length

Test Subject	Height (cm)	Average Arm Length (cm)
1		
2		
3		
4		
5		
6		
7		
8		

Source: Werner Williams

In Table 5, rearrange the data from Table 4 placing the data from the shortest person to the tallest person.

Table 5 Height and Arm Length from Shortest Test Subject to Tallest

Test Subject	Height (cm)	Average Arm Length (cm)
1(Shortest)		
2		
3		
4		
5		
6		
7		
8(Tallest)		

Source: Werner Williams

Average Heartbeats (beats/min.)

Sitting Standing Holding Breath

©AtthameeNi/Shutterstock.com

Average Time of Breath Holding (seconds)

©AtthameeNi/Shutterstock.com

After Rest After Exercise

Arm Length (cm)

©AtthameeNi/Shutterstock.com

Height (cm)

CONCLUSION

What is your conclusion? ____

____

Was your hypothesis supported by your study? ___________________________

QUESTIONS

1. What were some possible sources of error in the experiment you conducted? How would you make the experiment better to eliminate or reduce the possible sources of errors?

2. Experiments in science are usually reviewed by other Scientists. Why do you think this is done?

3. What is the difference between a hypothesis and a guess?

Lab 3: Food Analysis

All living organisms are composed of various types of organic molecules (molecules with carbons linked to other carbons) such as carbohydrates, protein, lipids and nucleic acids. Various indicator tests may be used to determine the presence of those substances in a food sample. Indicator chemicals can show if the substances are present in a food sample by changing to a particular color. In this lab exercise you will perform several of these tests. For each test performed, there will be a negative control (a substance with the nutrient absent so should give a negative color result) and a positive control (a substance with the nutrient present so should give a positive color result).

Benedict's Test for Simple Sugars

Benedict's reagent is used as a simple test for sugars. The test is primarily for monosaccharides. Disaccharides are made up of 2 monosaccharides linked together and polysaccharides are many sugars strung like beads on a string.

Benedict's reagent will be used to determine the presence of a sugar in various substances and to determine the relative amount of the sugar present. Record your results in the data table provided. Note that "results" mean what you actually see. The "conclusion" is what you deduce based on your results. The mixture will turn green if a trace of the monosaccharide sugar is present and yellow if a small amount of sugar is present. An orange color results if a moderate amount of a sugar is present and red if a large amount is present. For example, if a tube with a food substance and Benedict's reagent turns orange after heating, you would indicate in results "orange." Your conclusion would be "moderate." If the color in the tube does not change (remains blue), it indicates a negative test result.

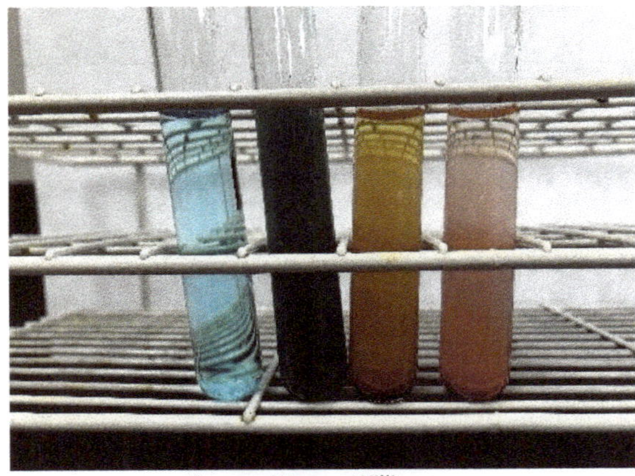

Source: Werner Williams

Blue-"Negative"; Green-"Trace"; Yellow-"Small"; Orange-"Moderate"; Red-""Large"

Materials

- Test tubes
- Test tube racks
- Test tube holder
- Hot water bath
- Marker
- Metric ruler
- Substances for testing
- Benedict's reagent

Procedure 1

1. Use two tubes and with a marker and a metric ruler, place a line on each test tube **1 cm from the bottom** of the tube. Number the tubes 1 and 2.
2. Fill test tube #1 up to the line with water and tube #2 with glucose.
3. Add **one dropper full** (not one drop) of Benedict's solution to each tube. Shake gently to mix the solutions.
4. Using a test tube holder, carefully place both tubes in the hot water bath and allow to heat for approximately 2 minutes or until a definite change is seen.
5. After heating, be sure to use the test tube holder to remove the tubes from the hot water bath. **Caution: tubes will be hot.**
6. Observe the colors of the tubes and complete the table below.

Test tube	**Substance**	**Results (colors)**	**Conclusion (amount)**
#1	Water		
#2	5% glucose		

Source: Werner Williams

Which tube serves as a negative control for simple sugars?

____

Which tube serves as a positive control for simple sugars?

____

Iodine Test for Starch

Iodine is used to test for the presence of starch. Long chains of glucose molecules may be arranged in plants to make starch molecules, which is a storage form of glucose. In the presence of starch, iodine turns from brownish-red to deep blue or blue-black.

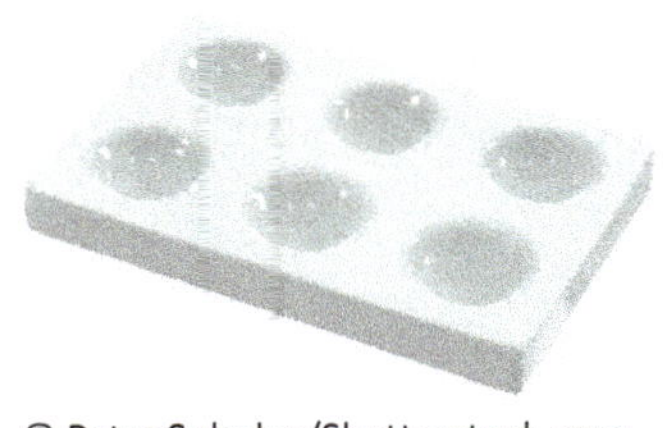
© Peter Sobolev/Shutterstock.com

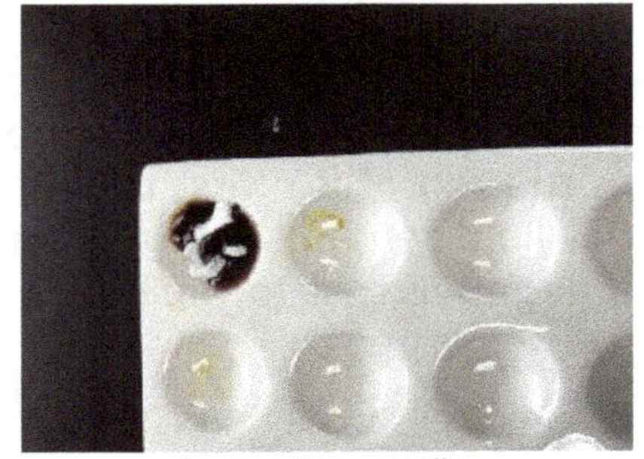
Source: Werner Williams

Spot Plates

Materials Spot plate
Iodine solution
Substances for testing

Procedure 2

1. Place a small amount of each of the substances listed in the table below on a spot plate.

2. Add 1-2 drops of iodine solution to each substance on the plate.

3. Note any color changes and record your data and conclusions in the table below.

Substance	**Results (colors)**	**Conclusion (starch present or not present)**
Water		
Cornstarch		
Glucose		

Source: Werner Williams

Which substance serves as a negative control for starch? ____________________

Which substance serves as a positive control for starch? ______________________

Sudan IV Test for Lipids

Lipids are molecules that are insoluble in water. Fats and oils are types of lipids in which a glycerol molecule is attached to one, two, or three fatty acids. These are called monoglycerides, diglycerides, or triglycerides.

Sudan IV stain will be used as the indicator stain. Following the procedure, a dark pink/red color on the top of the test tube indicates a positive test for lipids and the water beneath the lipid remains clear. A pale pink or no color result in the test tube would indicate a negative test for lipids.

Source: Werner Williams

Materials:

Test tubes
Test tube rack
Sudan IV stain
Marker
Metric ruler
Substances for testing

Procedure 3

1. Use two test tubes and number them 1 and 2 with a marker.
2. Using a metric ruler, place a line on each test tube **2 cm from the bottom** of the tube and another line on each test tube **4 cm from the bottom.**
3. Fill test tube #1 up to the 2 cm line with water and tube #2 up to the 2 cm line with salad oil.
4. Add water up to the 4 cm line in each test tube.
5. Add three drops of Sudan IV stain to each tube. Shake gently to mix the solutions.
6. Let the tubes sit for three minutes.
7. Observe the colors and record your results (color observations) and conclusions in the table on the next page.

Substance	**Results (colors)**	**Conclusion (lipid present or not present)**
Water		
Salad oil		

Source: Werner Williams

Which of the substances serves as a negative control for lipids? ________________

Which of the substances serves as a positive control for lipids? ________________

Biuret's Test for Protein

Amino acids are joined together by peptide bonds to form proteins. Proteins will react with biuret reagent and the mixture will turn from blue to purple.

Source: Werner Williams

Materials Spot plate
Biuret's solution
Toothpicks
Substances for testing

Procedure 4

1. Place a small amount of each of the two substances listed in the table on the next page, on a spot plate.

2. Add 1-2 drops of Biuret's solution to each substance on the plate. You may need to stir the mixtures with a toothpick.

3. Note any color changes and record your results in the table on the next page.

Substance	Results (colors)	Conclusion (protein present or not present)
Water		
Albumin (powdered egg white)		

Source: Werner Williams

Which of the substances serves as a negative control for proteins? _______________

Which of the substances serves as a positive control for proteins? _______________

Unknowns

You will be supplied with various foods to be tested. Preform each of the tests to the two foods that are assigned to your group. Record your results and conclusions in the data table below. Your group will be assigned two different foods to test but you are responsible for the results of foods tested by all lab groups.

Food	Simple Sugar		Starch		Lipid		Protein	
	Results	Con-clusion	Results	Con-clusion	Results	Con-clusion	Results	Con-clusion
1. Tuna								
2. Milk								
3. Peanut Butter								
4. Lettuce								
5. Refried Beans								

Source: Werner Williams

F. Food Labels

If you look at most food containers packaged for sale in the U.S, you will notice that nutritional information is included on the packaging label. This tool for evaluating packaged food products is required by the Food and Drug Administration (FDA). An example of one such label is seen below.

Nutrition Facts

Serving Size 5 oz. (144g)
Servings Per Container 4

Amount Per Serving

Calories 310	**Calories** from Fat 100
	% Daily Value*
Total Fat 15g	**21%**
Saturated Fat 2.6g	**17%**
Trans Fat 1g	
Cholesterol 118mg	**39%**
Sodium 560mg	**28%**
Total Carbohydrate 12g	**4%**
Dietary Fiber 1g	**4%**
Sugars 1g	
Protein 24g	
Vitamin A 1%	• **Vitamin C** 2%
Calcium 2%	• **Iron** 5%

*Percent Daily Values are based on a 2,000 calorie diet. Your daily values may be higher or lower depending on your calorie needs:

	Calories	2,000	2,500
Total Fat	Less Than	65g	80g
Saturated Fat	Less Than	20g	25g
Cholesterol	Less Than	300mg	300mg
Sodium	Less Than	2,400mg	2,400mg
Total Carbohydrate		300g	375g
Dietary Fiber		25g	30g

Calories per gram:
Fat 9 • Carbohydrate 4 • Protein 4

© Shelby Allison/Shutterstock.com

- All the information on the nutritional label is for **one serving**. Many packages contain more than one serving of a product.
- The nutrients that are listed are among those that are considered to be the most important for good health.
- You may use the percent daily values to compare different products and to tell if a serving of a particular food is high or low in a particular nutrient.

Use the food label above to answer the following questions:

1. How many calories would you consume if you ate the entire contents of this package? ______________________________

2. How many milligrams of sodium are in one serving of this product?

    ____

3. How many grams of fat are in one serving of this product? ________________
4. How many servings of this product would you need to eat in order to get the daily requirement of calcium? ______________________

Lab 4: Enzymes

Life would be impossible without enzymes. For example, without enzymes, it would take months or years for our muscles to accumulate enough energy for us to sit in our chairs. With enzymes present, the many reactions occurring in our bodies may take a fraction of a second.

Reactions in a cell take place in pathways in which starting materials called **substrates(s)** are converted into **products (p).**

$$A + B \rightarrow C + D$$

Substrates Products

These reactions may either combine simpler substrates in order to make complex products in a process called **anabolism** or the reactions may break down a complex substrate into simpler products in a process called **catabolism.** Fig.1

Enzymes (E), are proteins that speed up the transformation from substrates to products. The enzyme physically binds to the substrate and releases a product(s). Figure 1

$$E + S \rightarrow E\text{-}S \text{ complex} \rightarrow E + P$$

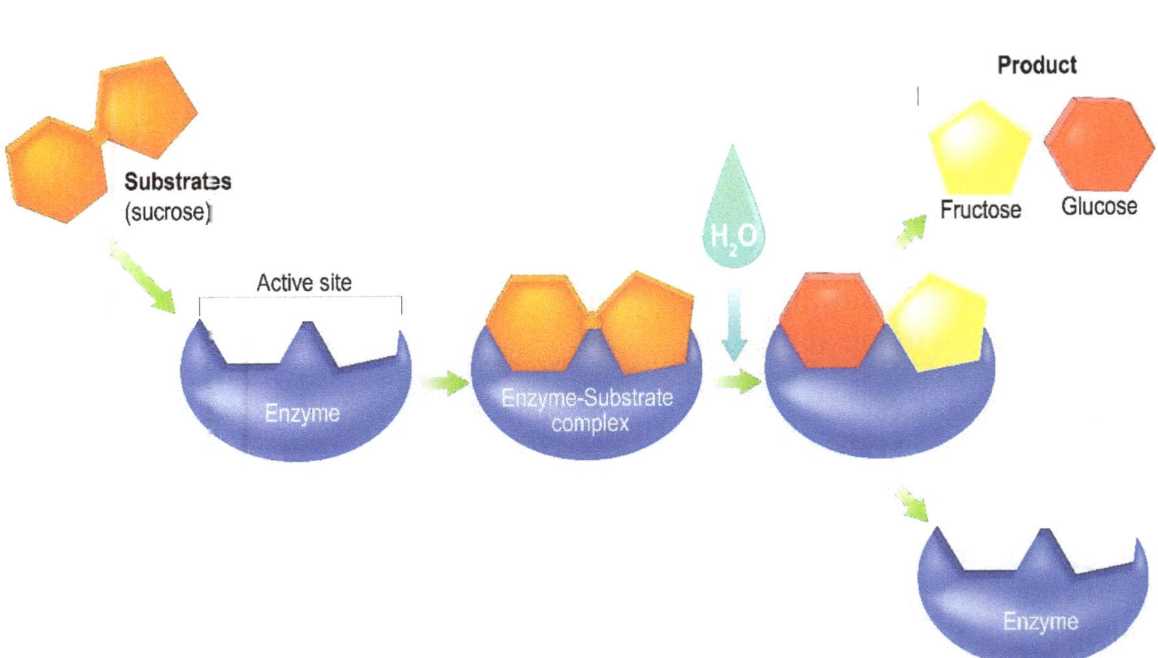

© Designua/Shutterstock.com

Figure 1 A Reaction Illustrating Catabolism

The part of the enzyme that bind to the substrate is called the **active site**. That site has a shape that is unique and binds to just one substrate or type of substrate. A cell only needs a small amount of an enzyme because since the enzyme retains its original shape when the reaction is completed, it may be used repeatedly to binc to more substrates. Some enzymes may bind to thousands of substrates/second.

To function well, an enzyme usually requires a closely controlled temperature and pH environment. Most enzymes like warmth and a pH that is close to neutral, but extremes in temperature or pH would generally denature an enzyme, which changes the enzyme's shape, preventing it from working. If we add more enzyme or substrate, there would be more collisions between the two, which means the enzyme should work faster so more product would be made.

In today's lab, we will test the effects of temperature, pH and enzyme concentration on an enzymatic reaction.

CATALASE

When cells use oxygen, they usually produce **hydrogen peroxide**, which is toxic (poisonous). In order to prevent the peroxide from doing damage, our cells also make an enzyme called **catalase**, which breaks down the hydrogen peroxide into harmless water and oxygen. The reaction is:

catalase

$$2\ H_2O_2 \rightarrow 2\ H_2O + O_2$$

hydrogen peroxide → water + oxygen

As oxygen is released, you will see bubbling. The faster the reaction proceeds, the more product will be made so the more bubbling you will observe.

<u>Procedure 1- Catalase Activity</u>

1. Mark 3 test tubes from the bottom at the 1cm and 3cm levels. Label the tubes 1, 2 and 3.
2. Fill <u>tube 1</u> to the 1st mark from the bottom with 1% catalase.
3. Add hydrogen peroxide to the 2nd mark. Shake gently to mix and wait a few seconds for bubbling.
4. Fill <u>tube 2</u> to the 1st mark from the bottom with water.
5. Add hydrogen peroxide I to the 2nd mark. Shake gently to mix and wait a few seconds for bubbling.
6. Fill <u>tube 3</u> to the 1st mark from the bottom with 1% catalase.

7. Add glucose solution to the 2nd mark. Shake gently to mix and wait a few seconds for bubbling.
8. Place your results in Table 1.

Use the following legend to indicate the degree of bubbling: 0- no bubbling; + little bubbling; ++ moderate bubbling; +++- much bubbling.

Table 1 Catalase Activity

Tube	Contents	Degree of bubbling	Explanation (Enzyme activity is present or absent)
1	Catalase + H_2O_2		
2	Water + H_2O_2		
3	Catalase + glucose solution		

Source: Werner Williams

What is the Effect of Temperature on Enzyme Activity?

Procedure 2- Effect of Temperature

1. Mark 3 test tubes at the 1cm and 3cm levels. Label the tubes 1, 2 and 3 and put your initials on each tube.
2. Add 1% catalase to each of the 3 tubes to the 1st mark from the bottom.
3. Place tube 1 in an ice bucket with ice (0°C), tube 2 in an incubator (37°C-body temp.), and tube 3 in a boiling water bath (100°C) for 15 minutes.
4. Remove the three tubes and then add hydrogen peroxide to each tube to the 2nd mark. **(Use cold hydrogen peroxide for tube #1).**
5. Shake gently to mix and wait a few seconds for bubbling.

   Place your results in Table 2.

Table 2 Effect of Temperature on Enzyme Activity

Tube	Temp.(°C)	Degree of Bubbling	Explanation
1	0		
2	37		
3	100		

Source: Werner Williams

What is your conclusion about the effect of temperature on enzyme activity?

____

____

If someone has a very high fever, what could be the effect on their enzyme activity?

____

____

What would be the effect of warmth on the movement of enzymes and substrates?

____

____

Would this increase enzyme activity?_______________

What is the Effect of pH on Enzyme Activity?

Procedure 3- Effect of pH

Sodium hydroxide (NaOH) and hydrochloric acid (HCl) are both strong and caustic bases and acids. Please be very careful in handling them. If they should spill on your skin, rinse immediately with water and notify your instructor.

1. Mark 3 test tubes at the 1cm, 3cm and 5cm levels. Label the tubes 1, 2 and 3.
2. Fill each tube to the first mark from the bottom with 1% catalase.
3. Place HCl (a strong acid) in tube 1 to the second mark.
4. Place pH 7 buffer solution in tube 2 to the second mark.
5. Place NaOH (a strong base) in tube 3 to the second mark.
6. Wait two minutes.
7. Add hydrogen peroxide to each of the tubes up to the third mark.
8. Shake gently to mix and wait a few seconds for bubbling.
9. Place your results in Table 3.

Table 3 Effect of pH on Enzyme Activity

Tube	pH	Degree of Bubbling	Explanation
1	Strong acid		
2	7(neutral)		
3	Strong base		

Source: Werner Williams

Catalase works best at which of the 3 pH's that you tested? ____________

What probably happened to the shape of the catalase when it was exposed to a strong acid or strong base?

____

Why does using the wrong pH usually decreases enzyme activity?

____

What is the Effect of Concentration on Enzyme Activity?

Procedure 4- Effect of Concentration

1. Mark two tubes at the 1 cm and 3 cm levels and label the tubes 1 & 2.
2. Fill test tube 1 to the 1st mark from the bottom with 1% catalase and test tube 2 with 0.1% catalase.
3. Add hydrogen peroxide to both tubes up to the 2nd mark.
4. Shake gently to mix and wait a few seconds for bubbling.
5. Place your results in Table 4.

Table 4 Effect of Enzyme Concentration on Enzyme Activity

Tube	Amount of Enzyme	Degree of Bubbling	Explanation
1	1%		
2	0.1%		

Source: Werner Williams

Would the results be the same if you allowed more time for the reaction to occur?

____

Would you get similar results if instead of changing the enzyme concentration, you changed the substrate concentration?
Explain____

____

____

Why does increasing the enzyme concentration increase enzyme activity?

____

____

Questions

1. What is the name of the part of the enzyme that binds to the substrate?__________________
2. What is the name of the type of reaction that builds complex molecules?_________________
3. What is the name of the type of reaction that breaks complex molecules into simpler ones?_________________________________
4. The targeted molecules of an enzyme are called what?
   ____
5. If we boil an enzyme, what happens to its activity?
   ____
6. If an enzyme is warmed, what happens to its activity?
   ____
7. If we add more enzyme, in the same amount of time, is the product made faster? ____
8. In the catalase reaction, bubbles are being given off. What gas causes the bubbling? ___________________________
9. If we create an unfavorable pH environment, what is most likely to happen to the enzyme activity?____
10. If we add an inhibitor that binds to the enzyme, what do you think would happen to the enzyme activity?
    ____
11. Name the starting materials of a reaction. __________________________
12. In the catalase reaction, what is (are) the substrate(s)?
    ____
13. In the catalase reaction, what enzyme is used?
    ____
14. In the catalase reaction, what products are made?_________________________

15. Are enzymes proteins, fats or carbohydrates?

____

16. What gas when used by a cell produces toxins like hydrogen peroxide?

____

17. Do you think an enzyme like catalase would work well in the stomach? Why or why not?____

18. Most enzymes in the body would work best at about what pH?

____

19. Is a pH of 3 an acid or a base?____________A pH of 11? ______________

20. Most enzymes in the body would work best at what temperature?

____

21. What usually happens to the shape of an enzyme when it is denatured?

____

22. If we have a high fever, why could this be deadly? _______________________

____

23. In an experiment, catalase was added to 4ml of H_2O_2. After 1 minute, 1 ml of H_2O_2 remained. How much ml's of H_2O_2 were consumed in the reaction? _______________If we repeated the experiment and found 2.5 ml of H_2O_2 remaining, would the second reaction have a higher or lower rate of reaction? __________

Lab 5: Human Genetics

Humans have 46 chromosomes in their body cells. 44 of them are called **autosomes** and the other 2 are called the **sex chromosomes**. Females have 2 X sex chromosomes and males have an X and a Y sex chromosome.

We have two genes for most of our biological characteristics, one gene from Dad and one from Mom. Letters are used to represent our genes, also called alleles, with a capital letter representing the dominant characteristics and a lower-case letter representing the recessive characteristic. If an individual's genotype is homozygous dominant (FF) or heterozygous (Ff), their phenotype will show the dominant characteristic. If an individual is recessive (ff), their phenotype will show the recessive characteristic.

Let's look at some common autosomal human traits. With your lab partner, determine each other's phenotypes from the characteristics listed in Table 1. Record your phenotype and probable genotype in the table. You may visualize these phenotypes on the next page.

Table 1 Autosomal Human Characteristics

Characteristic: d=dominant; r=recessive	**Possible Genotypes**	**Your Phenotype**	**Your Genotype**
Skin pigmentation: freckles (d); no freckles(r)	FF or Ff ff		
Hairline: widow's peak (d); straight hairline (r)	WW or Ww ww		
Bent little finger: little finger bends toward ring finger (d); straight little finger (r)	BB or Bb bb		
Thumb hyperextension: regular thumb (d); hitchhiker's thumb (r)	TT or Tt tt		
Earlobes: unattached (d); attached(r)	UU or Uu uu		

Source: Werner Williams

© Tracy Whiteside/Shutterstock.com

Freckles

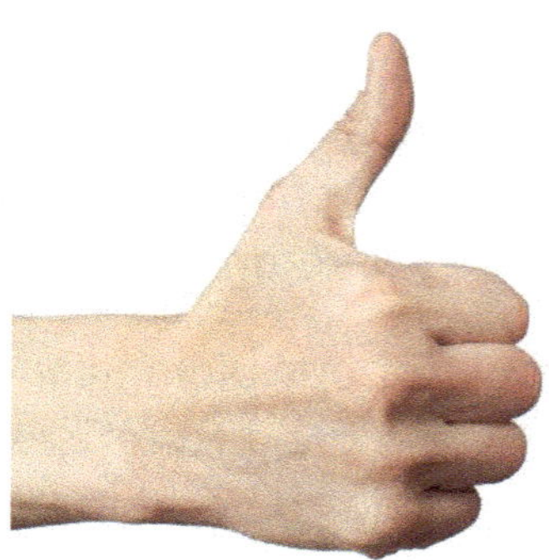

© Malyugin/Shutterstock.com

Regular thumb

© Elsa Hoffmann/Shutterstock.com

Hitchhiker's thumb

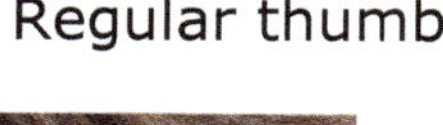

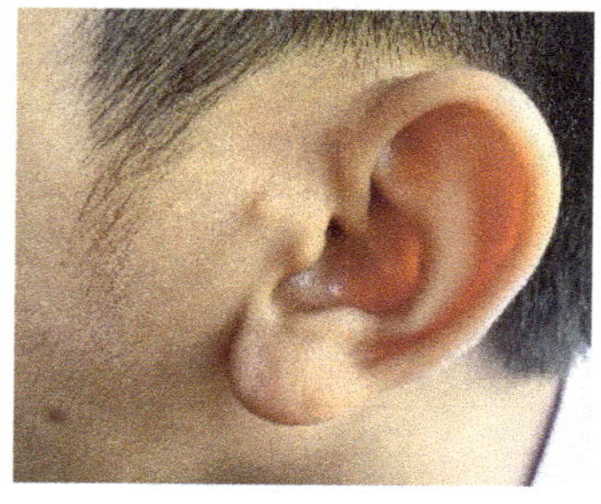

© Thiti Sukapan/Shutterstock.com

Unattached earlobes

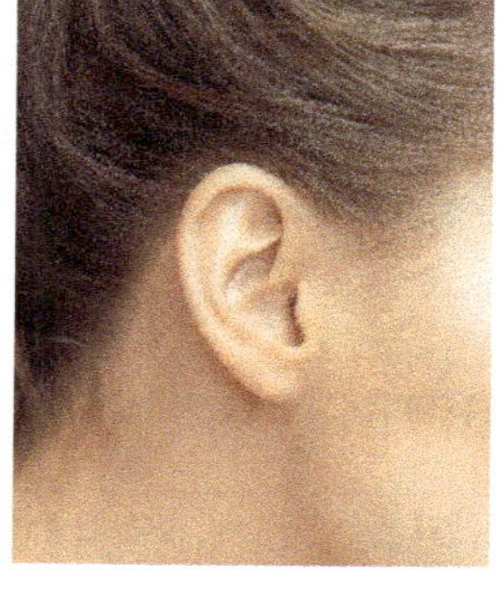

© Syda Productions/Shutterstock.com

Attached earlobes

© Everett Collection/Shutterstock.com

Widow's peak

© chombosan/Shutterstock.com

Straight hairline

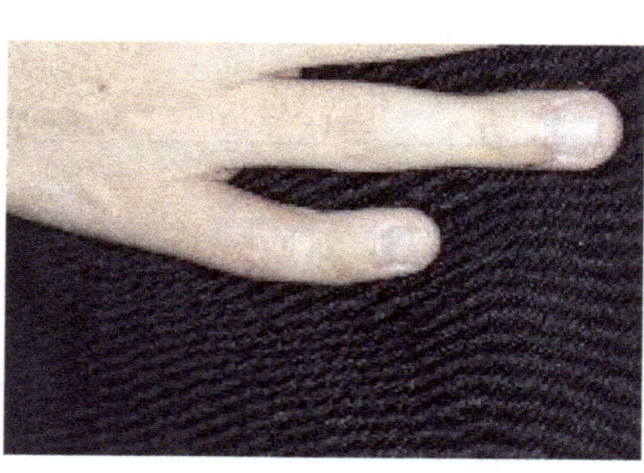

© Dr P. Marazzi / Science Source

Bent little finger

© Monster Ztudio/Shutterstock.com

Straight little finger

Genetics Problems

1. Mike's parents have freckles but Mike does not. What is the expected percentage among the children in this family? % with freckles________; % without freckles_____________

2. Susan is adopted. She has hitchhiker's thumb. Could both of her parents have hitchhiker's thumb? ________________ Could both of her parents not have hitchhiker's thumb?________Explain.

3. Janelle and everyone in her immediate family have straight hairlines. Her maternal grandmother has widow's peak. What is the genotype of her maternal grandmother? _____
Her maternal grandfather is no longer alive. What could have been his genotype?_________

Tasting PTC

Some people have the ability to taste PTC (phenylthiocarbamide) which is bitter, while others cannot taste it. Tasting (T) is dominant to non-tasting (t).

Procedure 1

1. Taste a piece of paper that contains PTC. Can you taste it?____________
2. What is your probable genotype (s)? ______________________________

Genetics Problem

Jordan's parents can taste PTC but Jordan cannot taste it. Construct a punnet square to show the probable genotypes and phenotypes of the possible children in that family.

ABO Blood Types

There are three genes that control the ABO blood types: A, B and O. A and B are dominant over O and A and B are co-dominant to each other. Each person has 2 of the 3 possible genes.

Table 2 Blood Types

Genotypes	Phenotypes (Blood Type)
AA or AO	A
BB or BO	B
AB	AB
OO	O

Source: Werner Williams

Genetics Problems

1. Could a man with blood type B and a woman with type AB produce a child with type O blood? ____________________

2. If the mother is blood type O and the father is A, what could be the blood type(s) of their children ?____

3. A boy wonders if he is adopted. If the father is blood type AB and the mother is blood type O, what blood type would indicate that the child might have been adopted? Blood type________________

Colorblindness

Colorblindness is an example of a sex-linked recessive characteristic. Sex-linked means the gene is found on one of the sex chromosomes (with color-blindness, it is found on the X chromosome). If the person has a recessive genotype, the person would be colorblind. Consider the following genotypes and phenotypes of various individuals:

Females

X^BX^B (normal vision)

X^BX^b (normal vision)

X^bX^b (color-blind)

Males

X^BY (normal vision)

X^bY (color-blind)

Genetics Problems

1. Rianna and her father are both color-blind but her mother has normal vision. What are the genotypes of these members of the family?
   Rianna__________; Dad________;Mom ______________

2. Rianna's sister Caroline has normal vision. What is her genotype?

   ____

3. The only colorblind member of Bob's family is his brother Jacob. What is Jacob's genotype?__________________What is Bob's father's genotype?

   ____

4. What is Bob's mother's genotype?_______________________If Bob's sister Mary has a colorblind son, what is Mary's genotype?______________________

Pedigrees

Pedigrees show the inheritance of a genetic disorder within a family and allows you to determine whether any particular person in that family has the gene for that disorder. Pedigrees show the person's phenotype. We can use this information to determine what are the chances that a particular couple would produce a normal or affected child.

(A carrier is a person who does not show the affected characteristic but may transmit the characteristic to their children).

Figure 1 shows you the symbols used in a typical pedigree chart.

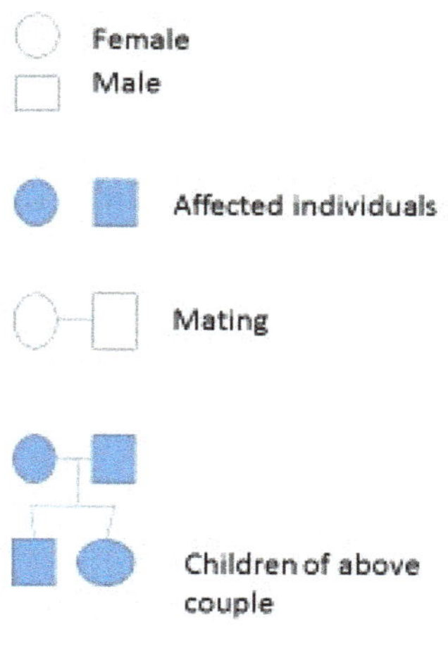

Source: Werner Williams

Figure 1 Pedigree Symbols

For each of the three pedigrees below, determine if the disorder or characteristic is autosomal dominant, autosomal recessive or X-linked recessive **and write the person's genotype beside their circle or square**. Use Table 3 to help you.

Table 3 Pedigree Solution

The Key That Solves The Pedigree	Possible Genotypes
Autosomal Dominant	AA or Aa= affected; aa= normal
Autosomal Recessive	AA or Aa= normal; aa= affected
X-linked Recessive	X^AX^A and X^AX^a = normal female; X^aX^a = affected female; X^AY = normal male; X^aY = affected male

Source: Werner Williams

Problems

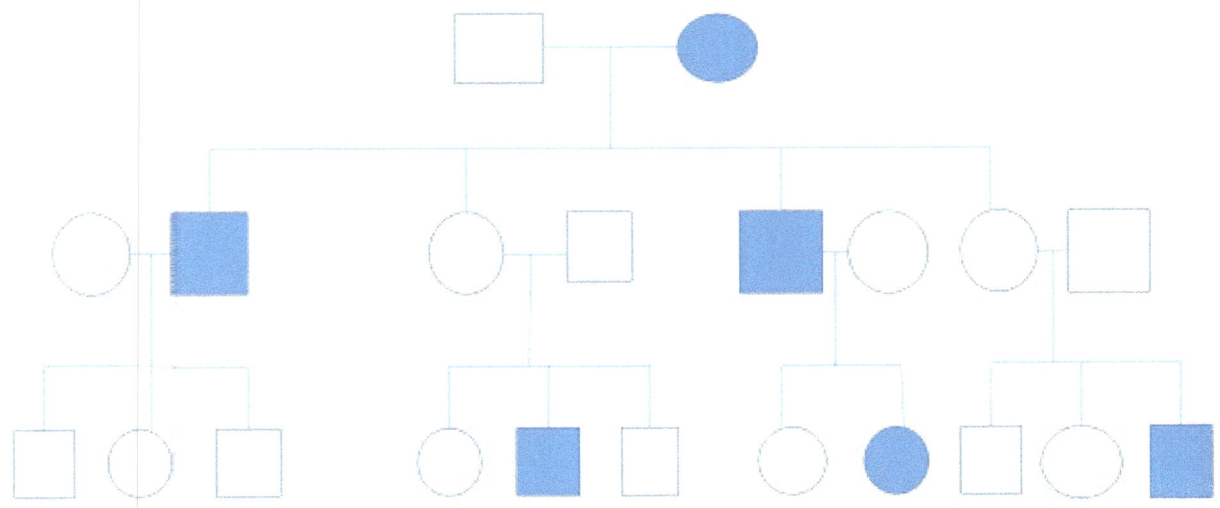

Source: Werner Williams

Key to this pedigree: ______________________________

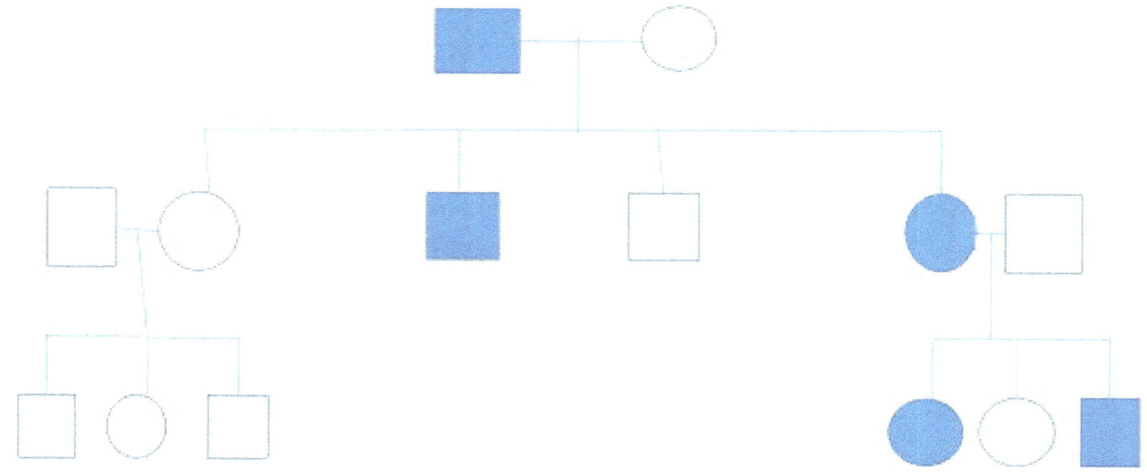

Source: Werner Williams

Key to this pedigree:____

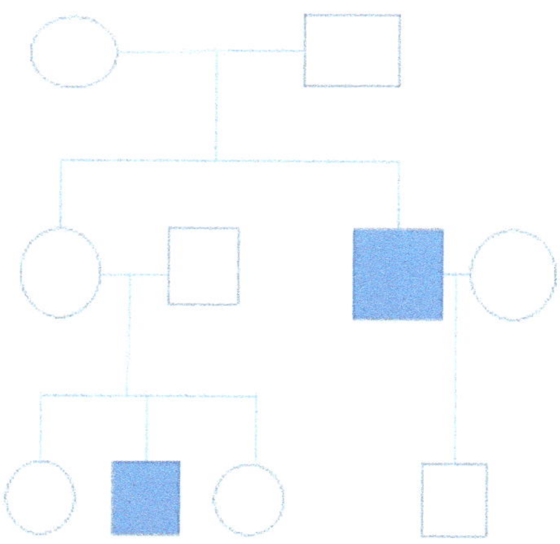

Source: Werner Williams

Key to this pedigree:____

Lab 6: Moving Materials Across Membranes (Diffusion & Osmosis)

Diffusion is the movement of materials from an area where there is a high concentration of those particles to an area where the concentration is lower. This process occurs in living and non-living systems and may continue until the particles are equally dispersed, called equilibrium.

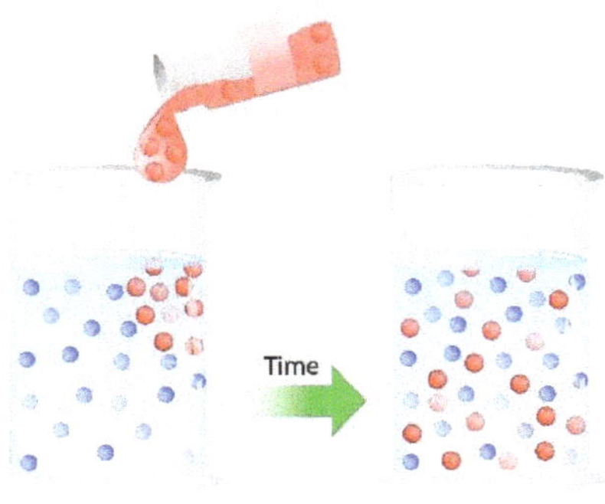

© Designua/Shutterstock.com

Diffusion

Cells can also use diffusion to bring food and other essential materials from the outside, through the cell membrane to the inside of the cell.

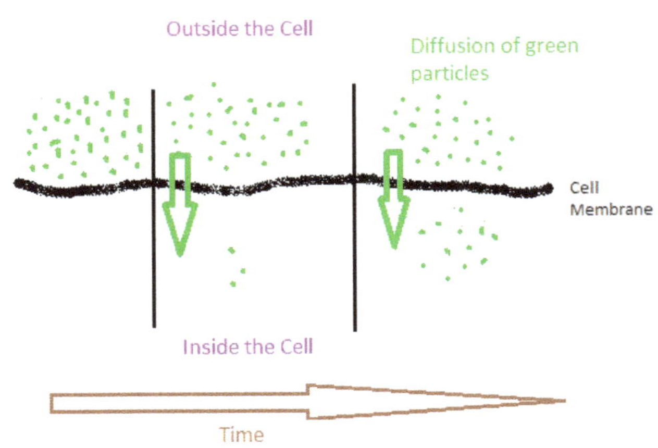

Source: Werner Williams

Osmosis is the special name given to the diffusion of water. This is based on at least two factors: 1) the water (called the solvent because it does the dissolving) moving from high to low concentration; 2) the concentration of the particles that are dissolved in the water called solutes. The solutes are attached to some water molecules, but the water molecules that are not attached are called "free." The concentration of the solutes determines the concentration of the "free" water in the solution.

Tonicity is the relative concentrations of the water and solutes on the outside and inside of a cell. Tonicity will have an effect on osmosis. Three types of solutions can be found outside of a cell:

1) **Hypotonic solution**- This solution outside the cell has a lower concentration of solutes than the solutes found in the cell. The solution will have more free water than the water found in the cell, so water will enter the cell and the cell will swell.

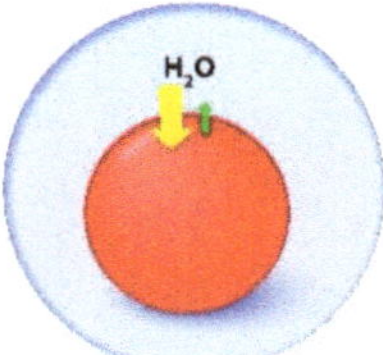

© Designua/Shutterstock.com. Modified by Werner Williams

2) **Hypertonic solution-**This solution outside the cell has a higher concentration of solutes than the solutes found in the cell. The solution will have less free water than the water found in the cell, so water will leave the cell and the cell will shrink.

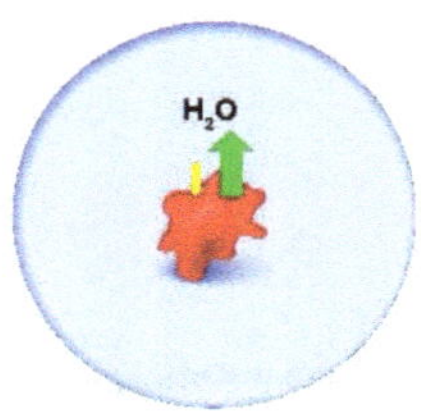

© Designua/Shutterstock.com. Modified by Werner Williams

3) **Isotonic solution**- This solution outside the cell has the same concentration of solutes as that found in the cell. The concentration of water outside and inside the cell are equal, so water will enter and exit the cell at the same rate. The cell's size will not change

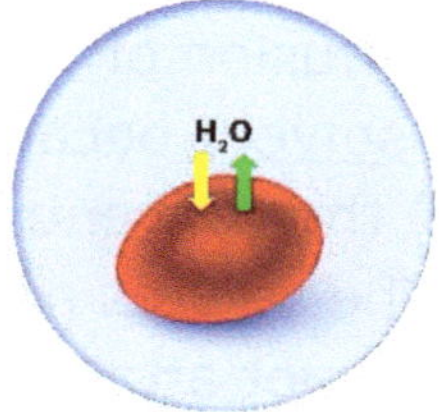

© Designua/Shutterstock.com. Modified by Werner Williams

In today's lab, two procedures will illustrate osmosis and diffusion.

Procedure 1 - Eggs and Osmosis

In this procedure, a hard-boiled egg will be left to soak either in water or a sucrose solution, and its weight over time will be measured as it either absorbs or loses water.

1) Pour 150 ml of the sucrose solution or water that is assigned to your group into a beaker.
2) Obtain a hard-boiled egg without its shell, and weigh the egg to the nearest 0.1g. Record this weight in Table 1 as the weight at 0 minutes.
3) Place the egg in the beaker with the sucrose solution or water.
4) After 35 minutes, remove the egg from the beaker, dry it with a paper towel and record the weight in Table 1.
5) Share your data with the other groups and complete Table 2.

Table 1 Weight of The Egg Over Time________________% Sucrose Solution

Time	Weight of the Egg (g)	Change in Weight (+ or -)
0 min		
35 min		

Source: Werner Williams

Table 2 Change in Weight For The Various Sucrose Solutions

Sucrose Solution	Change In Weight (+ or -)
0%	
10%	
20%	
30%	
40%	

Source: Werner Williams

Since the egg is about 75% water, would you expect the egg to gain or lose weight when placed in 100% water? Explain

____

____

Would you expect it to lose or gain water when placed in 55% water? Explain

____

____

Procedure 2

Starch is a storage form of glucose found in plants, glucose is a monosaccharide and iodine is an atom that contains 53 protons. In this procedure you will determine which of these three particles is large enough to move through the pores of a dialysis tubing.

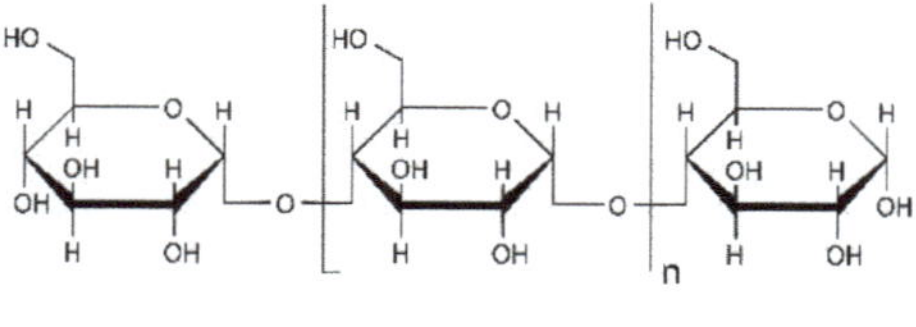

© chromatos/Shutterstock.com

Starch

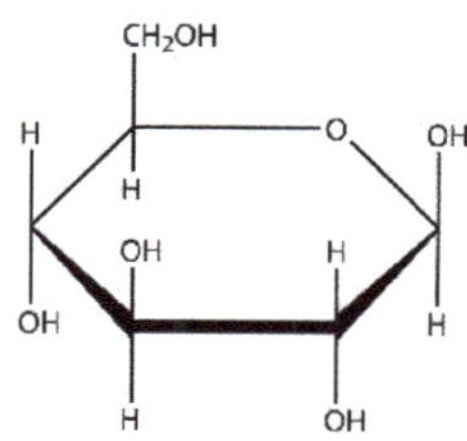

© chromatos/Shutterstock.com. Modified by Werner Williams

Glucose

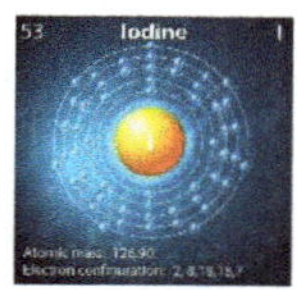

© BlueRingMedia/Shutterstock.com

Iodine

Procedure 2- Diffusion Into or Out of Dialysis Tubing

1. Obtain a beaker ½ filled with distilled water and add about 3 droppers-full (not drops) of iodine solution (enough to make the solution golden yellow). Record the color in Table 3.

2. Use a glucose test strip to test for the presence of glucose in the beaker. Dip the strip into the solution, remove immediately, and wait 30 seconds before comparing the color with the scale on the bottle. Record the presence or absence of glucose in Table 3.
3. Dip a new test strip into the beaker of the glucose/starch solution. Record the presence or absence of glucose in Table 3.
4. Obtain a pre-soaked dialysis tubing (bag) and 2 pieces of string.
5. Fold the bag over onto itself at one end and tie a **secure knot** at that end using one string. Leave about ½ inch between the string and the end of the bag.
6. Squeeze the other end of the bag between your fingers to open that end. Use the transfer pipet in the container of the glucose/starch solution to stir the solution and use the pipet to place the solution into the open end of the bag until it is about 1/2 full.

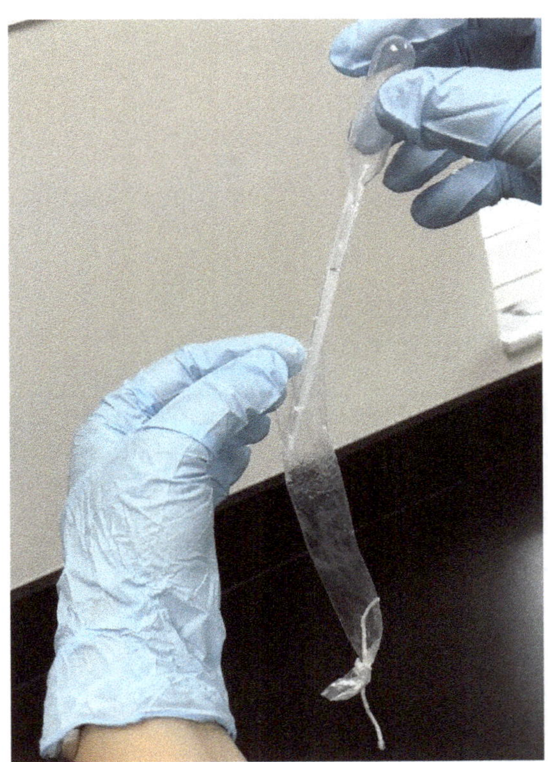

Source: Werner Williams

7. Squeeze the bag gently to remove excess air, fold the open end of the bag over onto itself and tie with string. The bag should not be so filled that it is tight or turgid. <u>Cut off the excess string.</u>
8. <u>Rinse the bag with tap water and blot dry with a paper towel.</u>
9. Record the color of the glucose/starch solution in the bag in Table 3.
10. Place the bag into the beaker of water and iodine.

Source: Werner Williams

11. Once the results are apparent (in about 20 minutes), record the color in both the beaker and the tube, then use a test strip to test for the presence of glucose in the beaker.

Table 3. Color Records of Liquid Contents

	Color of Water in Beaker after Iodine added	Presence of Glucose in Beaker Water	Presence of Glucose in Glucose/Starch Solution	Color of Glucose/Starch Solution in Dialysis Tube
Start				
After 20 minutes			Don't test	

Source: Werner Williams

1. What substance was small enough to exit the bag?
2. What substance was small enough to enter the bag?
3. What substance was too large to enter or exit the bag?

Lab 7: Fingerprinting

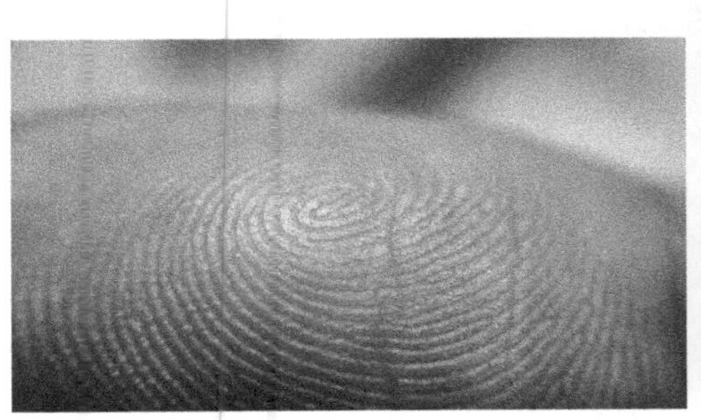

© Kritsana_MON/Shutterstock.com © TrifonenkoIvan/Shutterstock.com

Fingerprints are a commonly used way to identify people. Each finger has a set of ridges and valleys and is unique to each person. Even identical twins have different fingerprints. Fingerprints are difficult to change, usually last a lifetime and are used by police and the FBI to identify suspects in a crime, to identify people who are incapacitated or have died as a result of a natural disaster or to verify that people are who they say they are.

We develop fingerprints before we are born. Injuries like scrapes or burns do not change our pattern because when the skin grows back, it grows back with the same fingerprint pattern.

Fingerprints have been used for thousands of years for various reasons, but only recently did we develop the science of using it to reliably identify people. Fingerprint identification was first used in New York in 1902, when the New York Civil Service Commission fingerprinted applicants to prevent them from having more qualified applicants take their tests for them. The New York prison system a year later started to fingerprint criminals.

When we touch anything with pressure, we leave fingerprints. The oils and perspiration on our fingers, accumulate in the ridges, and when we touch an object, those liquids leave an invisible impression on that object. These latent prints are not visible to the naked eye, but by using powders or chemicals, they can be visualized. Patent prints are different and result when a substance on a finger touches a surface. For example, if a person touches paint and then presses their fingers to a wall, patent prints are left behind. Patent prints is more reliable than latent prints.

When the FBI attempts to identify someone by using their fingerprints, the first step is usually the use of a computer program that identifies possible matches in their fingerprint databases. Then human examiners take over and look at the shape of ridges, compare the points where the ridges start, end, split and join and scrutinize other points of similarity.

The Fingerprint Classification System has three basic fingerprint patterns: <u>whorls, loops and arches.</u>

© lestyan/Shutterstock.com © TrifonenkoIvan/Shutterstock.com © Kluva/Shutterstock.com

Whorl Pattern Loop Pattern Arch Pattern

It is useful to know the meaning of the following terms when analyzing fingerprints:

1. <u>Ridgelines</u>- The lines that make up a fingerprint.
2. <u>Core</u>-The approximate center of the fingerprint.
3. <u>Delta</u>- The area where two ridgelines come to a point.

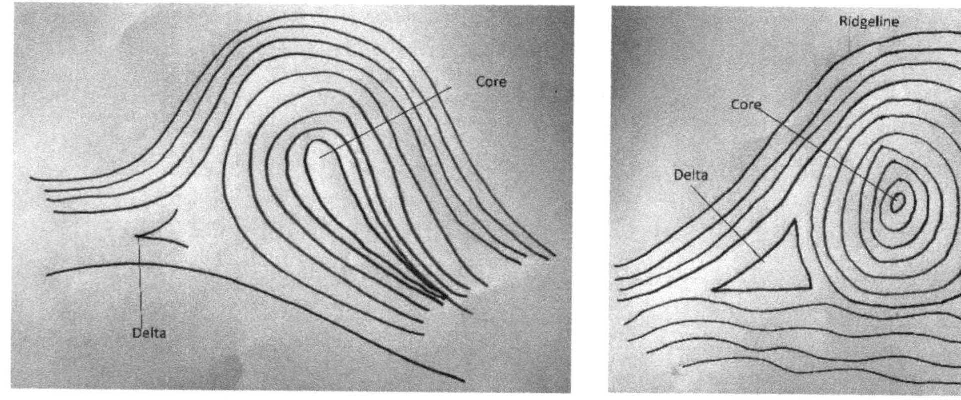

Source: Werner Williams Source: Werner Williams

4. <u>Whorl</u>- The complete circle that is formed by ridgelines between two deltas. Whorls do not continue off either side of the print.

© Kitch Bain/Shutterstock.com

Whorl

5. Arches: a. Simple Arch- Ridgelines enter on the left, rise, & exit on the right of the print.
   b. Tented Arch- Ridges meet at the center and form a peak.

© charobnica/Shutterstock.com

Simple Arch

© Leno4ek/Shutterstock.com

Tented Arch

6. Loops- Curved ridgelines that enter and exit on the same side of the print. Loops may face either left or right. A double loop has two separate loops on the same print.

© image bird/Shutterstock.com

Loop (Right Loop)

© leigh/Shutterstock.com

Double Loop

7. Mixed Prints- A fingerprint that does not easily fit into a loop, whorl or arch pattern. They show a combination of two or more basic patterns.

Procedure 1

Write the name of the fingerprint pattern (arch, whorl or loop) of the following three fingerprints:

© Yuri Samsonov/Shutterstock.com

© leigh/Shutterstock.com

© mighty chiwawa/Shutterstock.com

1.________________ 2.________________ 3.________________

Your Fingerprint Patterns

1. Obtain an inkpad, a magnifying glass and some paper towels.
2. You will be making a set of fingerprints by doing the following:
   a. Place one side of each finger on the inkpad and roll the finger from one side to the other, covering the front with ink but not the tip of the finger.
   b. Repeat the rolling motion to transfer a fingerprint to the appropriate square on this page. (Use a light touch. **Do not press too hard**).
3. After you have completed the process, wash your hands.
4. Write the name of the fingerprint pattern (arch, loop, whorl, mixed) under the print of each finger.

Left Hand

				Thumb

Right Hand

Thumb				

Lab 8: Blood Types and Typing

Blood typing involves the interaction between antigens and antibodies. **Antigens** are structures on the surface of cells that can help to identify them. **Antibodies** are proteins that cells produce in response to the presence of antigens.

Blood Groups

Red blood cells (rbc's) are covered with antigens. Two categories of rbc antigens are the ABO antigens and the Rh factor antigen. The ABO blood group system classifies blood into 4 groups: types A, B, AB and O.

If the surface of an rbc has the A antigen, the blood type is A.
If the surface of an rbc has the B antigen, the blood type is B.
If the surface of an rbc has both the A and B antigen, the blood type is AB.
If the surface of an rbc has neither the A nor the B antigen, the blood type is O.
Fig. 1.

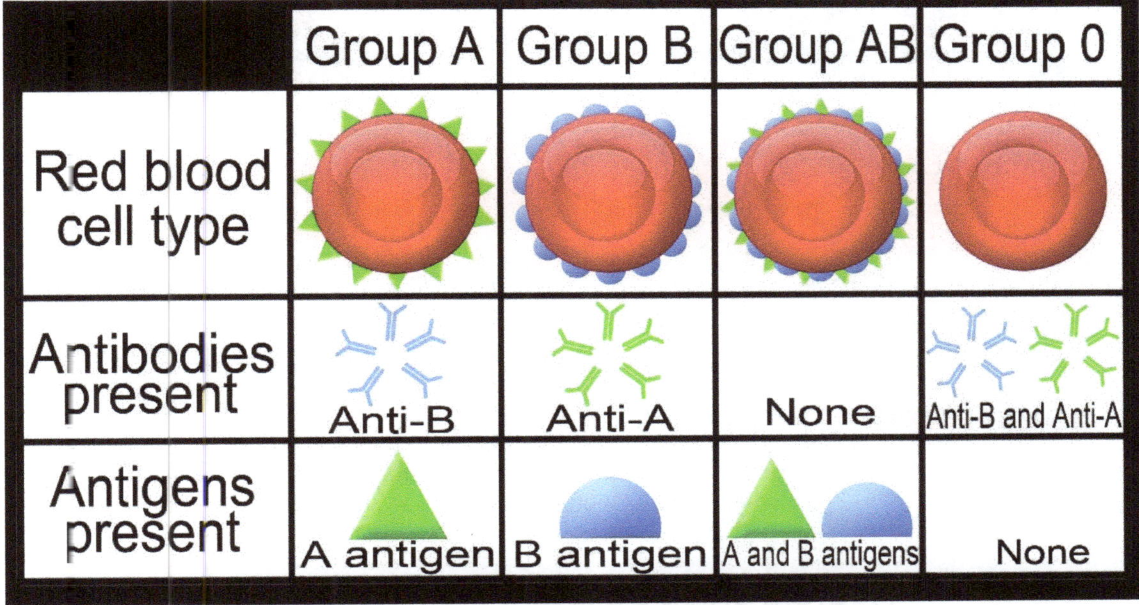

© ducu59us/Shutterstock.com

Figure 1 ABO Blood Group Antigens

Humans make antibodies against the antigens they lack, and the antibodies may destroy the antigens and the cells carrying them. Type A people make antibody B (anti-B), so they should not receive blood type B nor AB. Type B people make antibody A (anti-A), so they should not receive blood type A nor type AB. Type AB people do not make antibodies against blood type A nor type B, so type AB people may be given all blood types. Type O people make anti-A and anti-B antibodies, so should not receive blood types A, B nor AB. No antibodies are made against type O. Figure 1.

Type O blood may be given to anyone. They are the **universal donors**.
Type A people may give blood to type A or AB.
Type B people may give blood to type B or AB.
Type AB people may give blood only to type AB, but AB people may accept blood from anyone. They are the **universal acceptors**. Figure 2.

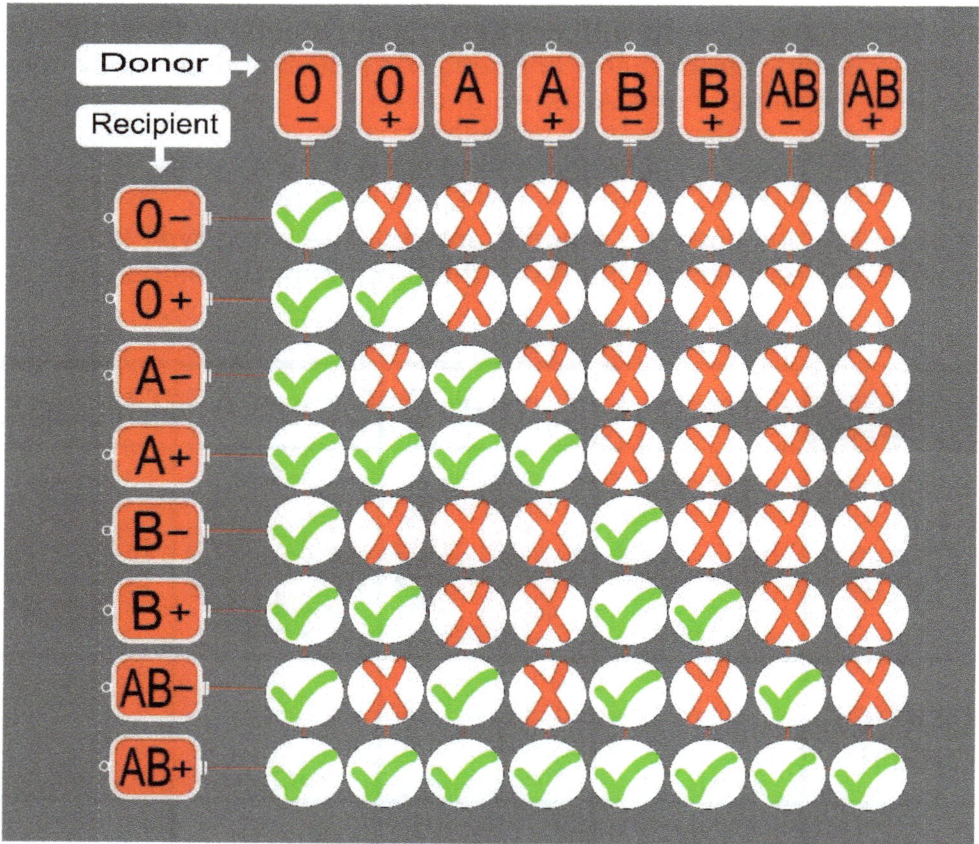

© hydra viridis/Shutterstock.com

Figure 2 Blood Donation Chart

Rbc's that have **the Rh_D antigen** on their surface are said to be **Rh positive** and those without the Rh_D antigen are said to be **Rh negative.** The Rh reaction is weaker than the ABO blood type reactions. Figure 3.

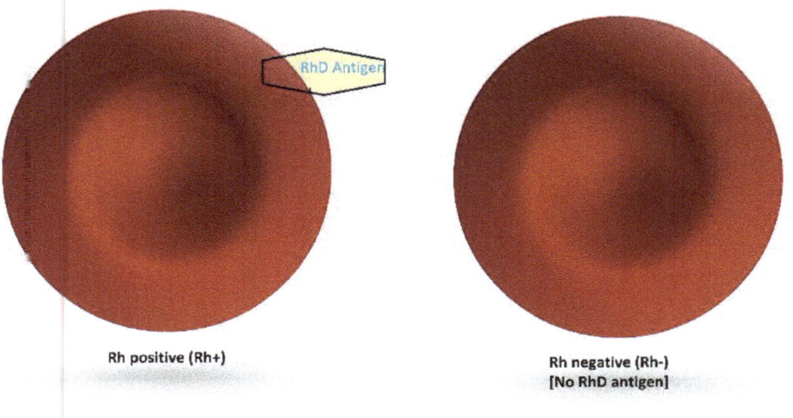

© Vitaly Art/Shutterstock.com. Modified by Werner Williams

Figure 3 Rh antigen

If a sample of blood is mixed with anti- A, anti- B and anti-Rh antibodies, and you notice clumping with anti-A only, the blood type of the sample is A negative. If the blood sample gives clumping with anti-B and anti- Rh only, the blood type of the sample is B positive. If the blood sample gives clumping with anti-A, anti-B and anti Rh, the sample is type AB positive and if it gives no clumping with anti-A, anti-B nor anti-Rh, it is type O negative. Figure 4 and Table 1

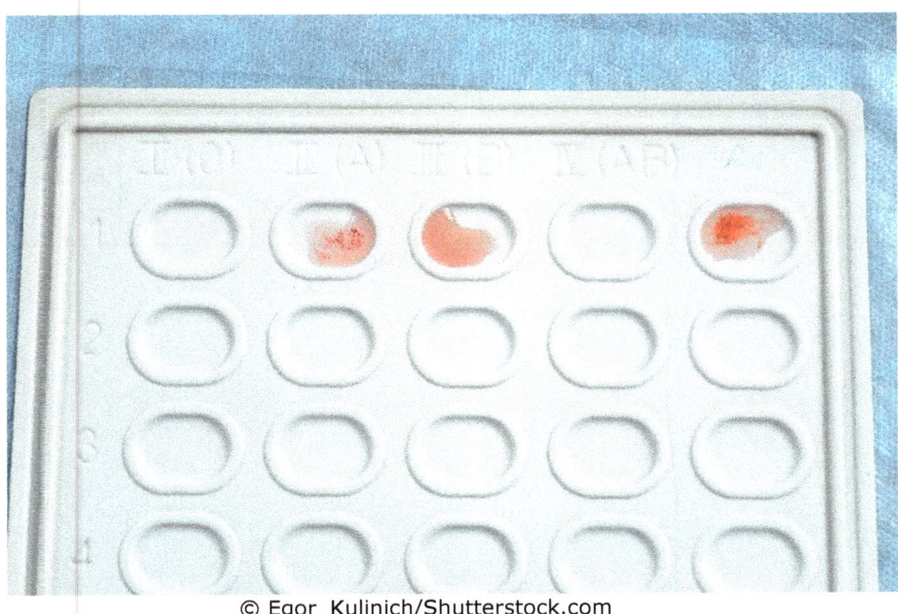

© Egor_Kulinich/Shutterstock.com

Figure 4 Interpreting Blood Types and Rh Reactions

Table 1 Interpreting Blood Types and Rh Reactions

Blood Type or Rh Reaction	Reaction when mixed with A antibody	Reaction when mixed with B antibody	Reaction when mixed with the Rh antibody
A	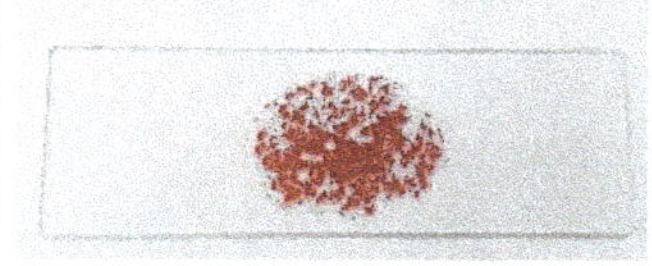 © Sunisa Butphet/Shutterstock.com	No reaction (No clumping)	
B	No reaction (No clumping)	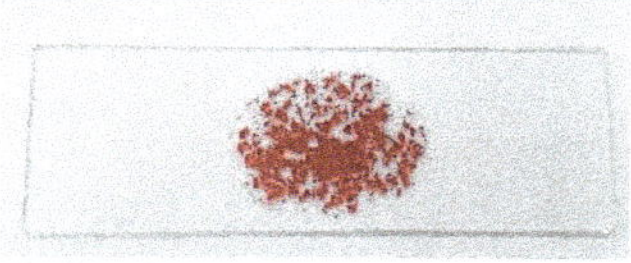 © Sunisa Butphet/Shutterstock.com	
AB	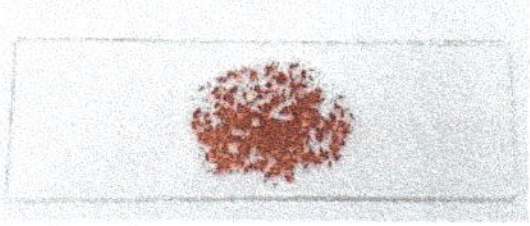 © Sunisa Butphet/Shutterstock.com	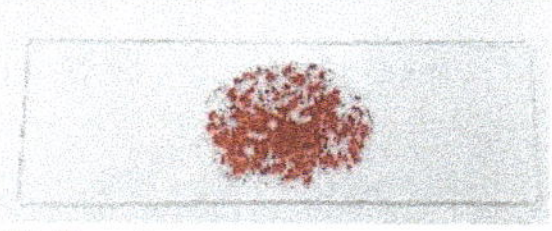 © Sunisa Butphet/Shutterstock.com	
O	No reaction (No clumping)	No reaction (No clumping)	
Rh +			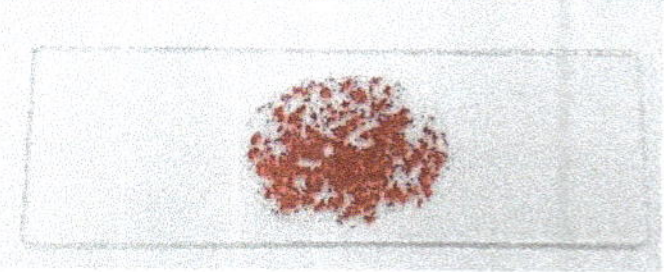 © Sunisa Butphet/Shutterstock.com

Source: Werner Williams

The proportion of the various blood types in the U.S. is shown in Figure 5.

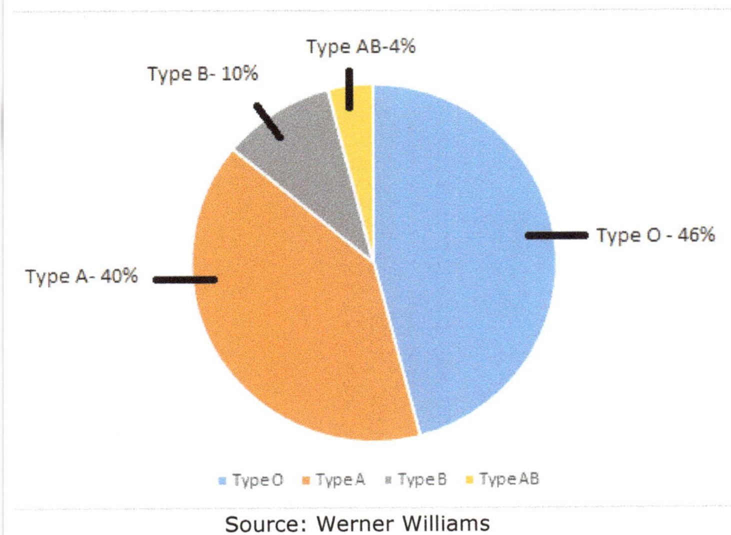

Source: Werner Williams

Figure 5 Proportion of Various Blood types in the U.S.

Questions

1. If a person is blood type A, what antibodies will they possess?
2. If a person is blood type B, what antibodies will they possess?
3. If a person is blood type AB, what antibodies will they possess?
4. If a person is blood type O, what antibodies will they possess?
5. If a person is blood type A, what blood types may they be given?
6. If a person is blood type B, what blood types may they be given?
7. If a person is blood type AB, what blood types may they be given?
8. If a person is blood type O, what blood types may they be given?
9. If you mix a sample of blood with anti-A and anti-B and anti-Rh antibodies, and you see clumping with anti-B only, what is the blood type and Rh reaction of that person?
10. If you mix a sample of blood with anti-A and anti-B and anti-Rh antibodies, and you see clumping only with the anti-Rh, what is the blood type and Rh reaction of that person?
11. If you mix a sample of blood with anti-A and anti-B and anti-Rh antibodies, and you see clumping with anti-A and anti Rh only, what is the blood type and Rh reaction of that person?

Lab 9: The Cardiovascular, Respiratory and Urinary Systems

Heartbeat

During a heartbeat, first the atria contract and then the ventricles contract. Usually there are two heart sounds with each heartbeat. The first sound (lub) is caused by the closing of the valves between the upper heart chambers and the lower chambers after the atria contract. The second sound (dub) is caused by the closing of the valves between the lower chambers and the arteries that drain the heart, after the ventricles contract.

Listening to the Heartbeat.

Procedure 1

1. You may listen to your own heartbeat or work with a partner. Obtain a stethoscope and place the earpieces in your ears. They should point forward.
2. Place the bell of the stethoscope on the left side of your chest or your partner's chest between the fourth and fifth ribs.
3. Which of the two sounds is louder, the lub or the dub? ________________
4. If you listened to your partner's heartbeat, switch and have your partner listen to your heartbeat.

The Pulse

When the heart contracts, it forces the blood into our arteries, and this stretches the arteries that exit the heart. The waves of blood exiting the heart as it contracts, creates the pressure that we call our **pulse**, which we can measure by probing the radial artery. Figure 1

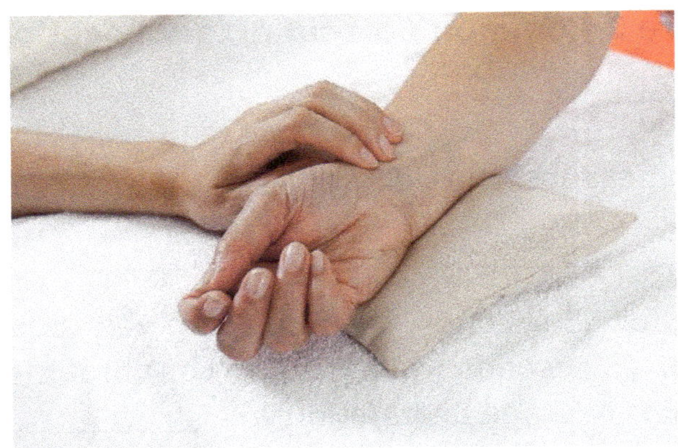

© Tyler Olson/Shutterstock.com

Figure 1 Radial Artery Location

Measuring the Effect of Exercise on Pulse Rate

Procedure 2

1. Find your pulse by placing your second and third fingers on the thumb side of your inner wrist. Press down slightly. Count your pulse for 15 seconds. What are the number of beats /15 seconds? ____________________
2. Multiply that number by 4 for the beats/minute. The number of beats/minute?__________________________________
3. Take your pulse again and average the two results. This is your average resting pulse rate. Average beats/minute?__________________________
4. Repeat #1, 2 and 3 above but this time measure your pulse at your common carotid artery. Figure 2. Average beats/minute?_______________________

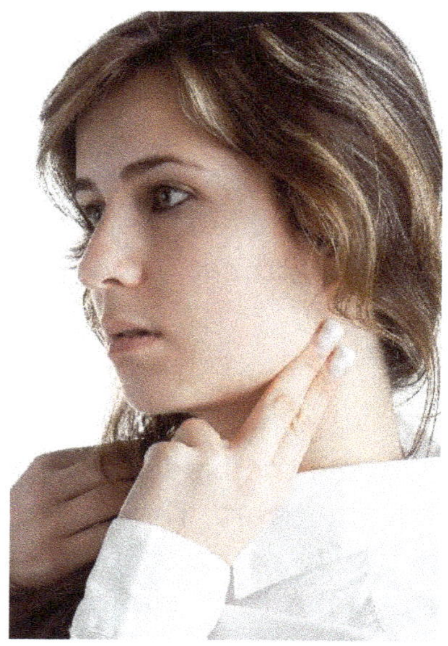

© ruigsantos/Shutterstock.com

Figure 2 Common Carotid Artery Location

How did the two measurements differ? ________________________

____

5. Hold your breath for 15 seconds, then measure your pulse using your radial artery for another 15 seconds while holding your breath. Multiply by 4 to get average beats/min. Average beats/minute?____________________

   Explain why holding your breath would affect your pulse rate. ___________

   ____

IF YOU HAVE HEART OR LUNG PROBLEMS DO NOT DO THE FOLLOWING EXERCISE!

6. Walk quickly down and up the stairs **twice** and then measure your pulse for 15 seconds. Multiply by 4 to determine the average beats/minute.

   Average beats/min?______________________________

   Explain why exercising affects your pulse rate.________________________

   ____

Blood Pressure

Blood pressure is the pressure that is exerted on the walls of blood vessels by blood, as it circulates through the arteries, veins and capillaries. When blood leaves the heart and enters the arteries, it moves with a pressure called the **systolic pressure**. When the heart relaxes, the arteries return to their original diameter as they squeeze blood into the rest of the cardiovascular system, and this pressure is called the **diastolic pressure**.

Blood pressure is measured in millimeters of mercury (mmHg), which is based on a device called a manometer, an inverted tube of liquid mercury. When pressure is placed on the mercury at the bottom of the tube, the mercury rises up the tube. The more pressure that is exerted at the base, the higher the mercury rises. So a pressure of 120mmHg will raise the mercury 120mm. More modern devices do not use mercury. Figures 3, 4.

We usually take blood pressure readings in the brachial artery, just above the elbow. The average systolic reading is about 120mmHg and the average diastolic would be 80mmHg. The reading is usually reported as the systolic reading/diastolic reading, so the average would be 120/80 ("120 over 80"). Blood pressure may be affected by certain factors like salt intake, age, volume of blood and temperature.

The normal range is about 90-140/60-90.

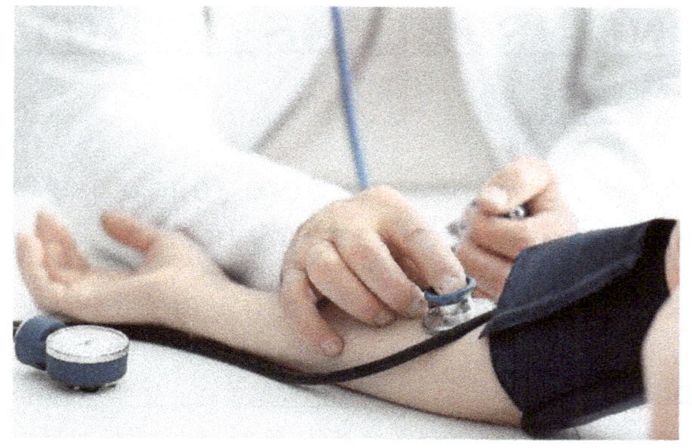

© Africa Studio/Shutterstock.com

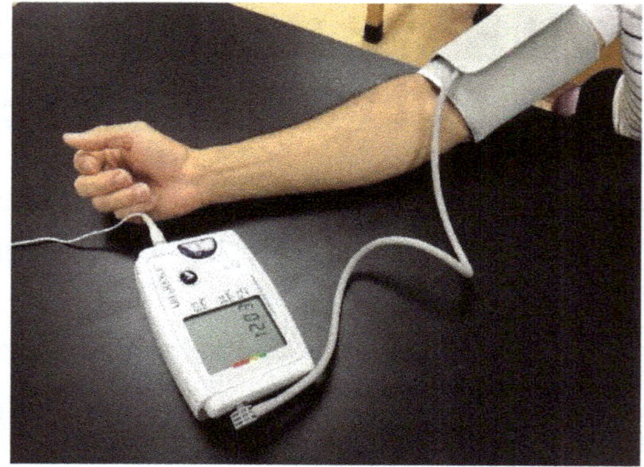

Source: Werner Williams

Figures 3 and 4. Measuring Blood Pressure

Measuring Blood Pressure

Procedure 3

1. Sit comfortably with your left arm resting on a flat surface so that the center of your upper arm is at the same height as your heart.

2. Lay your left arm on a table with your palm up. Place the Easy-Fit cuff on your upper arm. Align the white marker on the cuff over the brachial artery on the inside of the arm. The tube from the cuff should be facing downward and toward the inside of your arm. Fasten the cuff securely. The bottom of the cuff should be about 1" above your elbow.

3. The cuff should be snug but not too tight. You should be able to insert two fingers between the cuff and your arm.

4. Press the START button. As the cuff pressurizes, the measurement will begin. It is normal for the cuff to feel very tight.

NOTE: If an appropriate pressure is not obtained, the device automatically starts to inflate again.

NOTE: If you wish to stop inflation at any time, press the START button again.

5. When the measurement is complete, the systolic and diastolic pressure readings and pulse rate are displayed. The cuff deflates and the monitor automatically shuts off after 45 seconds, or you can turn it off by pressing the START button. Record the reading in Table 1.

6. Remove the cuff.

7. Have your partner lie down and relax for about 2 minutes. Attach the cuff and take the person's blood pressure while they are lying down. Record the reading in Table 1.

8. Have your partner stand up and record the pressure after 10seconds and then record the pressure again 5 minutes after rising, while the person is standing.

Table 1 The Effect of Position on Blood Pressure

Position	Blood Pressure (Systolic/Diastolic)
Sitting	
Lying Down	
Standing (10 sec. after rising)	
Standing (5 min. after rising)	

Source: Werner Williams

Did your blood pressure change after rising? Why? ______________________

____

Breathing

Normal breathing is controlled by a part of the brain stem called the respiratory center. It sends messages to various muscles that cause breathing movements. The rate and depth of breathing may be affected by certain factors. These include our emotional state and the presence in the blood of various chemicals like oxygen and carbon dioxide. For example, if the CO_2 concentration in the blood increases, our breathing rate increases.

Factors Affecting Breathing

Procedure 4

1. *Normal breathing*. To determine a person's normal breathing rate and depth, have your partner sit quietly first for 1 minute. Then ask the person to count quietly backwards, beginning with 300. While the person is counting, watch their chest movements and count the number of breaths in 1 minute. Take note of the depth of their breathing. Record the number of breaths in Table 2 as their normal breathing rate and also record the depth of their breathing as well.

2. *Effect of hyperventilation.* Have your partner sit and **guard them to prevent them from possibly falling over.** Have your partner breathe deeply and rapidly for 2 minutes. **If the person starts to feel dizzy, stop the procedure immediately**. After the hyperventilation period, count the number of breaths the person took in one minute and also note and record their breathing depth in Table 2.
3. *Effect of Holding Breath.* Have your partner sit quietly until their breathing returns to normal. Have the person hold their breath for as long as possible. As the person begins to breathe again, count the number of breaths the person took in one minute and also note and record their breathing depth in Table 2.

Table 2 Breathing Rate and Depth

Factor	Breathing Rate (breaths/min.)	Breathing Depth (shallow, medium, deep)
Normal		
Hyperventilation		
Holding Breath		

Source: Werner Williams

Urinalysis

Urinalysis may be a useful tool to help in diagnosing certain conditions. The test can easily be performed by dipping a urinalysis test strip into urine, and noting the color of the various square sections that would react with chemicals in the urine. Figure 5.

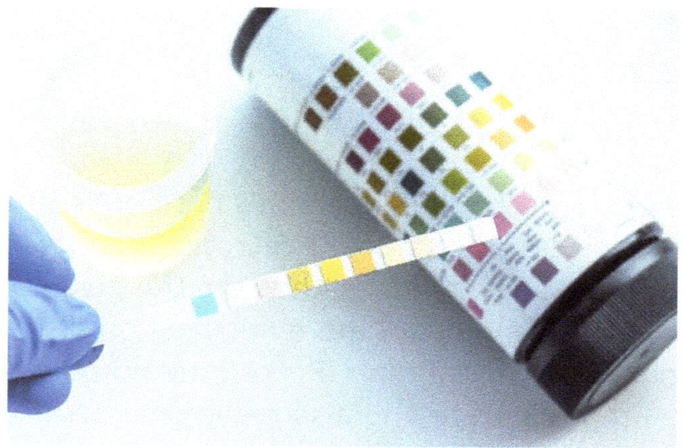

© Sirirat/Shutterstock.com

Figure 5. Urinalysis Test Strip Measuring System

Detecting Various Chemicals in Urine

Procedure 5

1. Obtain a test strip and a container with simulated urine. Imagine that this was from a patient who was complaining of excessive urination and thirst and feelings of extreme tiredness.
2. Dip the test strip into the urine (make sure all of the square sections are immersed under the surface of the urine) for no longer than 5 seconds.
3. Draw the edge of the strip along the rim of the specimen container to remove excess urine.
4. Turn the strip on its side, and tap once on a paper towel to remove any remaining urine.
5. After 1 minute, read the test strip by holding the strip next to the color blocks on the diagnostic color chart. Make sure that the strip is matched closely to the various sections of the chart.

   Is the glucose section within the normal range? ____________________

   Is the ketones section within the normal range? ____________________

   What condition might the person have? ____________________________

   If the person's urine contains glucose, why would the person feel tired?

   ____

OPTIONAL

6. Obtain an unused cup and a test strip, go to the rest room and urinate into the cup.
7. Repeat steps # 2-5.
8. Dispose of the cup and its contents in the rest room.
9. After returning to the lab, compare the color of the test strip to the diagnostic color chart.
10. Were there any abnormal readings?

    ____

Urinalysis Analysis (For general reference only)

Abnormal readings in one or more areas of a test strip may or may not indicate a serious condition. Some conditions that may result in abnormal readings in one or more areas of a urinalysis test strip are as follows:

Urobilinogen: Raised levels may be due to: cirrhosis, hepatitis, hepatic necrosis, hemolytic and pernicious anemia, malaria.

Protein: May be an early sign of kidney disease, or injury to the urinary tract, bladder or urethra, inflammation, malignancies, multiple myeloma.

pH :Low pH (acidic 6.0-6.6): Foods such as acidic fruits (cranberries) can lower the pH, as can a high protein diet. As urine generally reflects the blood pH, metabolic or respiratory acidosis can make it more acidic. Other causes of acidic urine include diabetes, diarrhea and starvation.

High pH (alkaline 7.5-8.5): Low carb or vegetarian diet. May be associated with kidney stones or urinary tract infection.

Specific Gravity (SG): Decreased SG may be due to: excessive fluid intake, renal failure or diabetes.

Increased SG may be due to: dehydration due to poor fluid intake, vomiting or diarrhea, heart failure, liver failure, inappropriate antidiuretic hormone secretion or diabetes.

Ketones: Presence of ketones may indicate diabetes, alcoholism, a state of starvation or pregnancy.

Bilirubin: Presence of bilirubin in the urine may indicate: liver disease, liver infection or pancreatic causes of jaundice.

Glucose: Glucose is not normally present in the urine. High amounts in the urine may indicate diabetes, liver disease, medications such as tetracycline, lithium, penicillin, cephalosporin or pregnancy.

Nitrite: Higher than normal amounts may indicate a urinary tract infection.

Blood: Higher than normal amounts may indicate renal disease, bladder or renal tumor, trauma to the kidneys or the use of anticoagulants.

Lab 10: The Senses

Our sense organs include the eyes, ears, tongue, nose and skin. These are associated with vision, hearing, tasting, smelling, touch, pressure, temperature and pain.

Stimuli from the outside of the body is received by these organs, changed into sensory impulses and then sent to the brain and/or spinal cord. In today's exercise, we will examine various aspects of some of our senses. `

Eye and Vision

The eye is a very complex organ that captures light rays and forms images. Those images are sent along the optic nerve to the brain. The structure of the eye is shown below as well as an explanation of the function of some parts of the eye:
Cornea: The tissue covering the eye.
Pupil: The opening to the lens.
Iris: The colored part of the eye.
Lens: The structure that focuses light rays onto the retina.
Retina: The back of the eye which has cells called cones for detecting colors in bright light, and the rods for seeing in dim light.

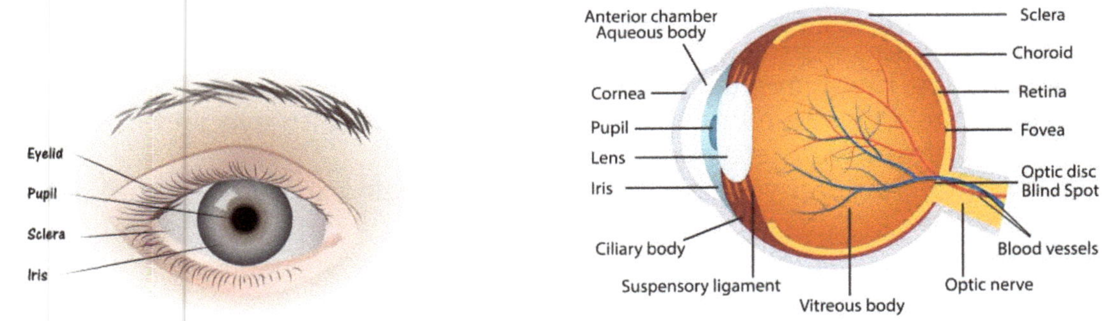

© pablofdezr/Shutterstock.com. Modified by Werner Williams © Neokryuger/Shutterstock.com

Eye (frontal view) Eye (side view)

Peripheral (Side)Vision

Our peripheral vision helps us see things out of the corner of our eye. It is a large portion of our entire visual field. The visual field is all we can see at one time without turning our head.

Procedure 1

1. Face your partner sitting at your desktop about two feet apart at eye level. Obtain a pencil with an eraser.

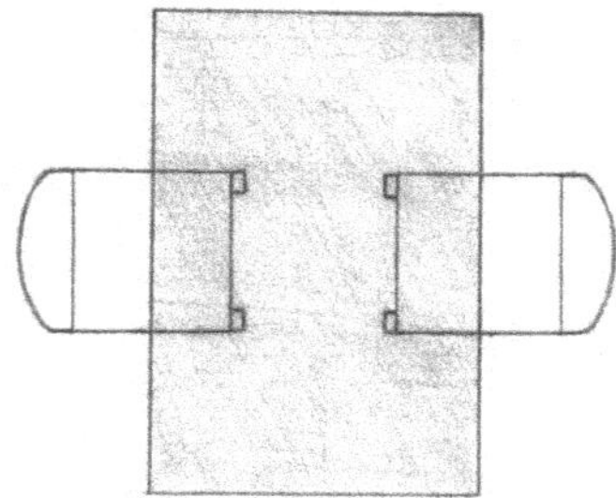

Source: Werner Williams

2. Have your partner cover one eye and you cover one eye also. For example, your partner covers his or her right eye and you cover your left eye.
3. Look directly into each others' open eyes without moving.
4. Now, with the pencil in your right hand, extend the eraser as far to your right as you can reach and slowly bring the eraser in towards you and your partner, keeping it on an imaginary line midway between the two of you. You will need to bend your elbow to do this. Both you and your partner should see the eraser come into view out of the corners of your eyes at about the same time.
5. If one of you sees the eraser much sooner than the other, the person who sees it last may have decreased peripheral vision.
6. Repeat the procedure but this time close the opposite eye and bring the eraser in from your left side.
7. Each time, you and your partner should be seeing the eraser at about the same time.

Depth Perception

Depth perception is the ability to see the world in three dimensions. It is important for us to determine how far an object is away from us. When we are walking, and we see an obstacle, we need to make sure that we perceive its distance from us correctly, so as not to walk into it. In this lab, we will explore how depth perception changes when we are using one eye compared to two eyes. Depth perception should be easier with both eyes open because each eye looks at the image from a different angle.

Procedure 2

1. Hold two pencils, one in each hand. Hold them either vertically or horizontally facing each other at arms-length from your body.
2. With one eye closed, try to touch the ends of the pencils together.

3. Repeat #2 above with both eyes open. It should be easier with both eyes open.

Visual Acuity

Visual acuity is the sharpness of vision. It is generally tested with a Snellen eye chart, which consists of letters of various sizes printed on a white card. This test is based on the fact that letters of a certain size can be seen clearly by eyes with normal vision at a specific distance. The distance at which the normal eye can read a line of letters, is printed at the end of that line.

Procedure 3

1. Stand 20 feet in front of a posted Snellen eye chart. There is a strip of masking tape on the lab floor that marks the 20 foot distance.
2. Have your lab partner stand at the Snellen eye chart. Cover one eye with your hand and read each consecutive line on the chart aloud and have your lab partner check for accuracy. If you wear glasses, take the test twice – first with glasses off and then with your glasses on. Do not remove contact lenses, but note that they were in place during the test.
3. Record the number of the line with the smallest-sized letters read correctly. If it is 20/20, your vision for that eye is normal. If the second number is more than 20, you have less than the normal visual acuity. If it is 20/40, you can see at 20 feet what a person with normal vision can see at 40 feet. If it is 20/60, you can see at 20 feet what a person with normal vision can see at 60 feet, and so on. If the visual acuity is 20/15, your vision is better than normal. You can see at 20 feet what a person with normal vision can see at 15 feet.
4. Repeat the above process with the other eye and record the results below.

Visual acuity, right eye without glasses ______________________________
Visual acuity, left eye without glasses ________________________________
If you normally wear glasses, record the results below:
Visual acuity, right eye with glasses _________________________________
Visual acuity, left eye with glasses__________________________________

Ear and Hearing

The ear is divided into three parts: the outer, middle and inner ear. The ear is responsible for converting sound waves into vibrations that can be changed into impulses (messages). Our equilibrium and balance are also regulated by the inner ear. The structure of the ear is shown on the next page as well as the function of some of the parts.

The Outer Ear

The pinna: Picks up sound waves
The ear(auditory) canal: Sends sound waves to the eardrum

The Middle Ear

<u>The eardrum</u>: vibrates when struck by sound waves.
<u>Three small bones</u>: bones that vibrate and transit the vibration to the cochlea.

Inner Ear

<u>The cochlea</u>: snail-like structure that converts waves into nerve impulses.
<u>The auditory (cochlea) nerve:</u> carries impulses to the brain.
<u>The semicircular canals:</u> help maintain balance and equilibrium.

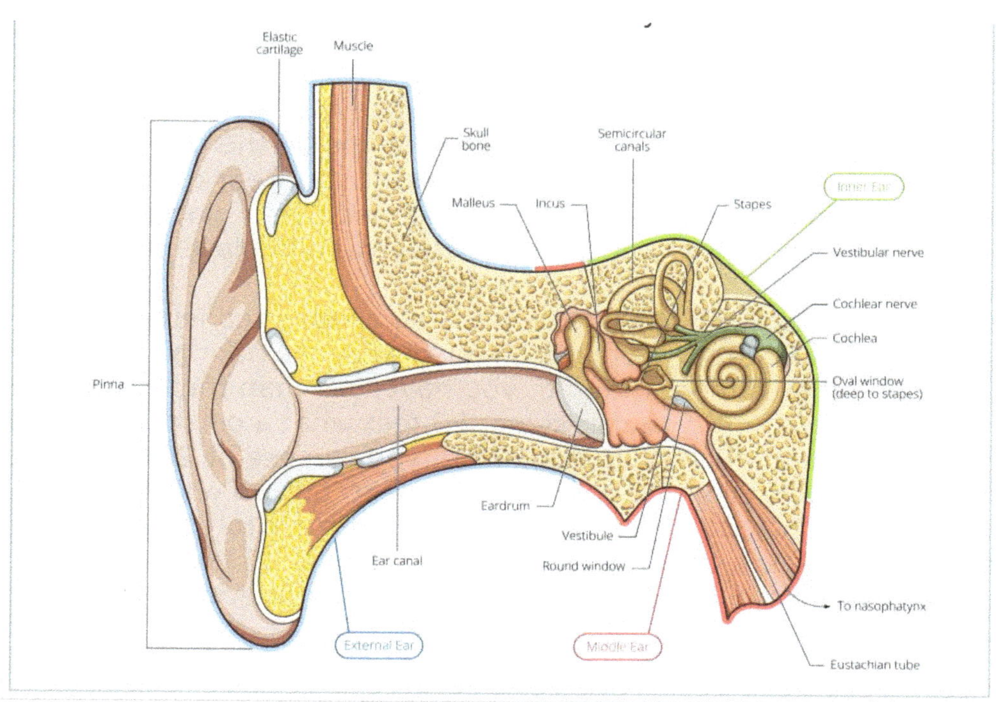

© Alexander_P/Shutterstock.com

Locating Sound

We locate the direction of sound according to how fast it is detected by either or both ears. If there is a difference in the hearing ability of our two ears, we can make a mistake about the direction of sound. Both you and your partner should perform the procedure on each other.

<u>Procedure 4</u>

1. Ask your partner to be seated with eyes closed.
2. Strike a tuning fork at the five locations listed in # 4 below. Choose the different locations randomly.
3. Ask your partner to give the exact location of the sound in relation to his or her head by pointing in that direction.
4. Record your partner's perceptions when the sound is:

a. directly below and behind the head ______________________________

b. directly behind the head____

c. directly above the head ____

d. directly in front of the face_______________________________________

e. to the side of the head ____

5. Change roles and repeat.

6. Is there an obvious difference in hearing between your two ears?
____

Hearing Tests

Weber Test

The Weber test is a test for determining a hearing impairment. It can distinguish between conductive hearing loss (a problem with sound waves moving from the outer to the inner ear) and sensorineural hearing loss (a problem with the cochlea or auditory nerve).

Procedure 5

1. Sit in a chair and have your lab partner strike a tuning fork, and while it is vibrating, your partner should touch the handle of the tuning fork to the top of your head.

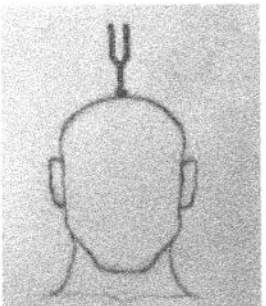

Source: Werner Williams

2. Is the sound the same in both ears or is it louder in the left or right ear?____
3. Change roles and repeat.

Test Results:
Normal hearing – sound is the same in both ears
Conductive hearing loss – sound is louder in the ear with conductive hearing loss.
Sensorineural hearing loss – sound is louder in the ear without hearing loss

Rinne Test
The Rinne test is a hearing test that compares hearing through the air and bone conduction hearing through the mastoid process.

Procedure 6

1. Sit in a chair and have your lab partner strike a tuning fork and while it is vibrating, your partner should hold the handle of the tuning fork on your mastoid process for a few seconds. After the sound diminishes, hold the fork a few centimeters from your ear.

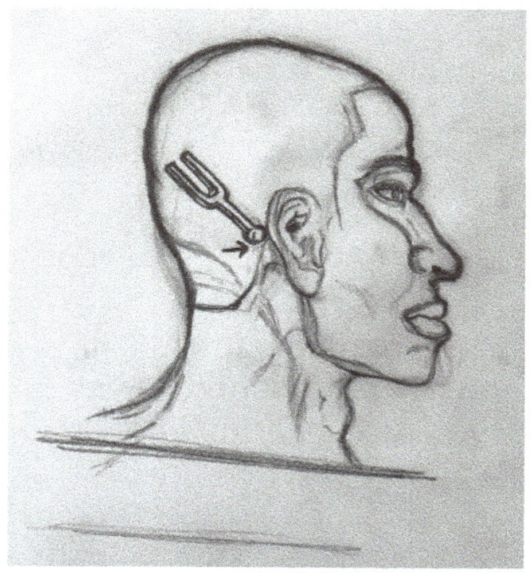

Source: Werner Williams

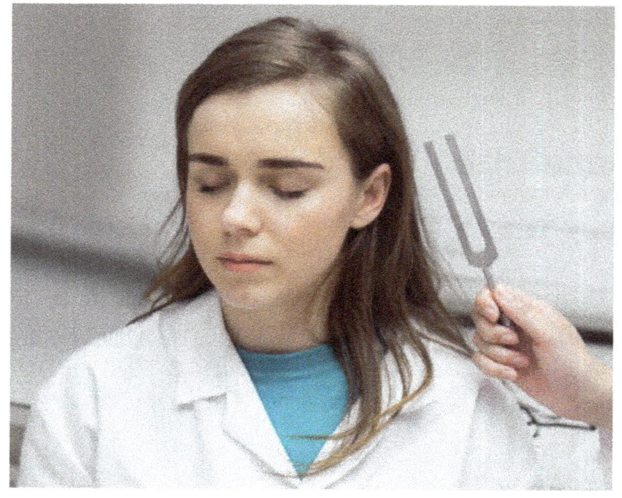

© wavebreakmedia/Shutterstock.com

2. Which way was the sound louder?
   ____________through the mastoid process__________through the air

3. Change roles and repeat.

Test Results:

Normal hearing – sound is louder through the air than through the mastoid process.
Conductive hearing loss – sound is louder through the mastoid process than the air.

Maintaining Equilibrium

Our sense of balance is dependent on input not just from the inner ear but also from information that a part of the brain called the cerebellum is assessing from our eyes and from other parts of the body.

Procedure 7

1. Work in pairs. Have one person stand on one foot with his or her eyes open and time how long he or she can maintain their balance. Set a limit of 2 minutes.
2. Repeat this activity with your partner's eyes closed, **standing by, to hold your partner if he or she loses their balance**.

   Record your results in terms of length of time:

   Eyes open ______________________________

   Eyes closed ______________________________

3. Change roles and repeat.

4. Explain your results
   ____
   ____

Taste and Smell

There are at least five major taste sensations: **sweet, sour, salty, bitter** and **umami**, which is a savory or meaty taste. MSG (monosodium glutamate) produces a strong umami taste.

We can use taste and smell to detect chemicals in our environment. Molecules in the air and food will stimulate not only cells in the nasal cavity but also cells in the taste buds on the tongue. Taste and smell are closely related partly because molecules can be easily transported between the mouth and the nose. That's why when we have a cold and our sinuses may be clogged, food tastes differently.

Connection Between Taste and Smell

Procedure 8

1. Have your partner close their eyes, hold their nose tightly, and then use a plastic spoon to give your partner a jellybean or skittle.
2. They should chew the candy for about 10 seconds and try to identify the flavor. Then have the person let go of their nose and continue chewing the candy and try again to identify the flavor.
3. Record the results.

Actual Flavor	Flavor While Holding Nose	Flavor After Releasing Nose

Source: Werner Williams

Touch

Two Point Discrimination

Different parts of the skin have different amounts of touch receptors. The two-point discrimination test measures the sensitivity of an area on the skin. It tests the ability to distinguish between two points.

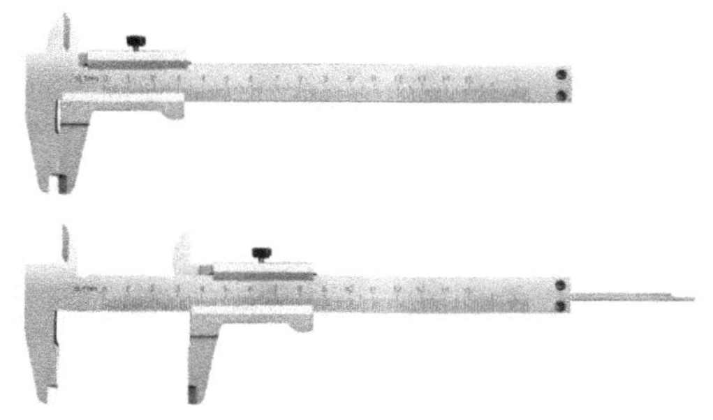

© para_graph/Shutterstock.com

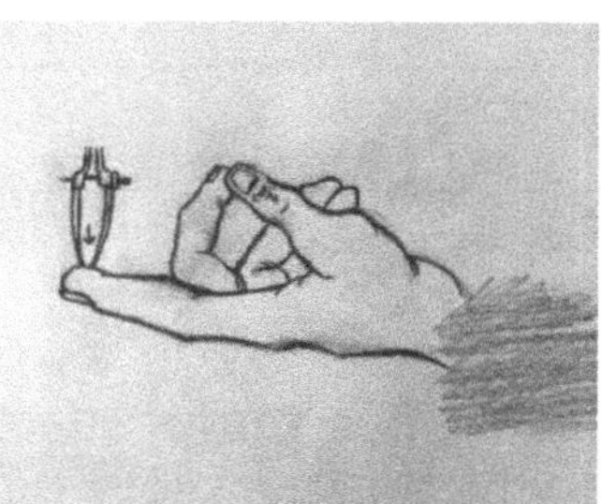

Source: Werner Williams

Procedure 9

1. Using a caliper, test your ability to differentiate between two distinct sensations when the skin is touched simultaneously by two points.
2. Close your eyes and have your partner test the areas of your body listed in the table below.
3. Starting the test with the caliper arms completely together, you should feel only ONE point touch you. Have your partner gradually increase the distance between the two arms, touching the same area of skin each time, until you feel TWO distinct points touch you. This measurement, the smallest distance at which you could distinguish two separate points of contact, is the two point discrimination threshold. (Your partner should randomly touch you with just one point to discourage you from guessing.)

4. As soon as you feel TWO distinct points touch you, have your partner measure the distance between the two caliper arms and record the results in the table below.

Body Area Tested	Two-point Discrimination Threshold (mm)
Back of hand	
Palm of hand	
Fingertips	
Back of neck	
Posterior of forearm	

Source: Werner Williams

Lab 11: Microorganisms

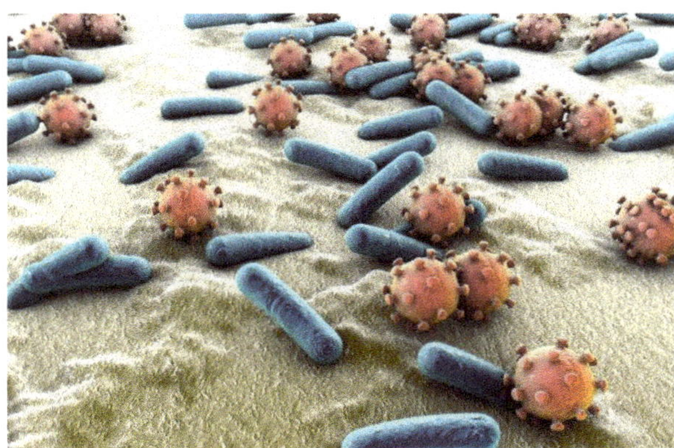

© Kateryna Kon/Shutterstock.com

Microbes are everywhere. They are found in the foods we consume, the air we breathe and the objects we touch. Even though a few may cause disease, microbes are necessary for our survival. They not only produce 50% of the air we breathe, but they fertilize the soil and decompose dead plants and animals. They are also used to make foods such as bread, yogurt, cheese, butter, Nutrasweet ®, pickles and chocolate. Many microbes even live in and on us, and actually contribute to our good health.

Their shape and arrangement helps in their identification. Most bacteria are either **round shaped** called **cocci** or **rod shaped** called **bacilli**. Others may be **curved shaped** called **vibrios** or **spiral shaped** called **spirilla or spirochetes**. All bacteria are single celled but some form chains or clusters of cells. If two round cells join together to form a short chain, we call them a **diplococcus**. Cocci that form a longer chain are called **streptococcus**, and **streptobacillus** is a chain of rod-shaped cells. A cluster of round bacteria is called **staphylococcus**.

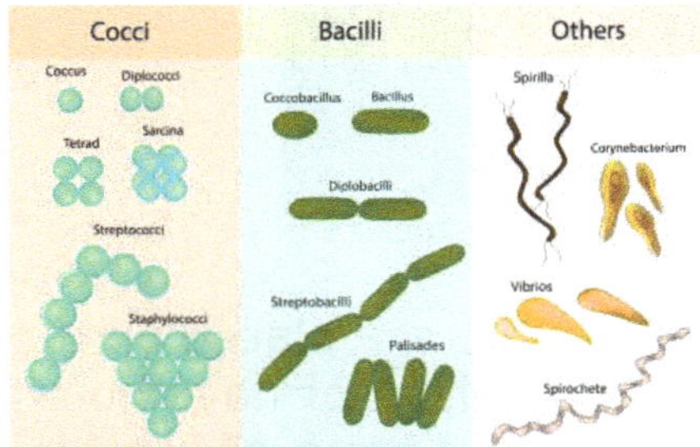

© Sakurra/Shutterstock.com

Some human diseases caused by bacteria may be treated with antibiotics. These drugs may damage or destroy certain microbial structures, such as their ribosomes or cell membranes. Since not all microbes have the same structures, by grouping them based on their structures, we can more easily determine if a particular antibiotic will affect it. Therefore, correctly identifying the microbe is key to the physician prescribing the appropriate antibiotic.

A common test used to identify microbes is called the **Gram stain**. It places many microbes into one of two groups based on a type of covering called the cell wall they possess. If a microbe has a thick layer of **peptidoglycan** (a combination of carbohydrates and proteins) in its cell wall, it is said to be **gram positive**. Gram positive cell walls bind a purple dye and at the end of the gram staining process, they appear **purple** under a microscope.

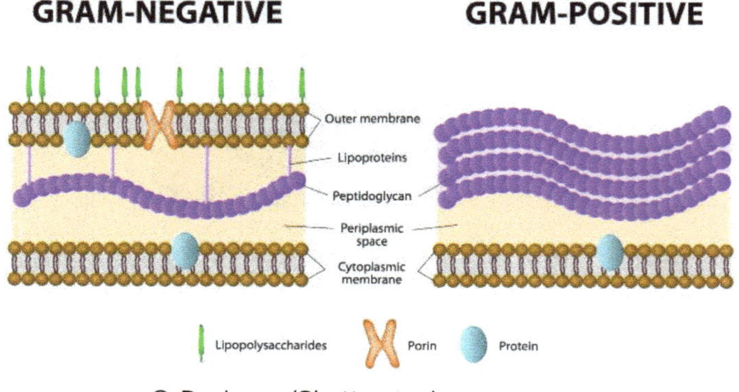

© Designua/Shutterstock.com

Other bacteria have a thin layer of peptidoglycan and after gram staining, they appear **red** or **pink**. Those cells are called **gram-negative**. Antibiotics like penicillin prevent some microbe from making peptidoglycan. Gram positive cells are more easily damaged by penicillin.

One of the ways to determine if an antibiotic will affect a microbe is to grow the microbe on food called media in a petri plate.

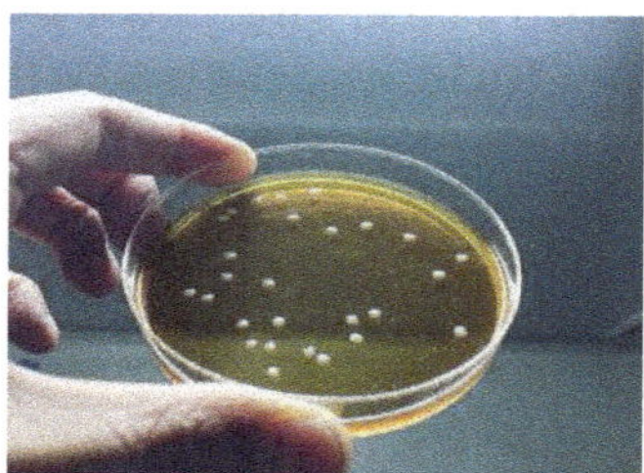

© KuLouKu/Shutterstock.com

Paper disks containing various antibiotics are placed on the plate. The drug in the disk will diffuse outward and may kill, inhibit or have no effect on the microbe. If it kills or inhibits the microbe, it will leave a **zone of inhibition** around the disk. That drug may be used to treat a patient with an infection that is caused by that microbe.

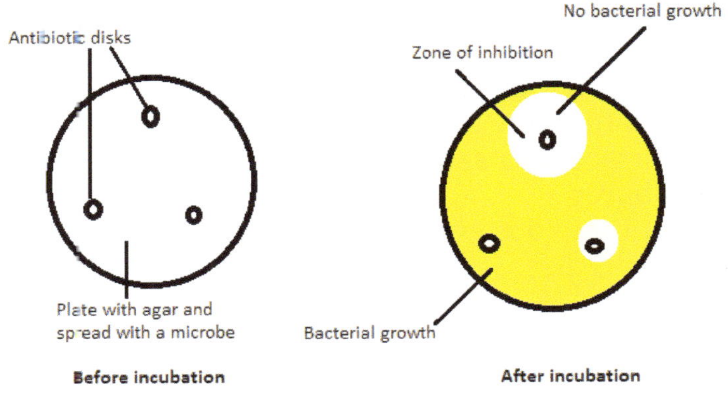

Source: Werner Williams

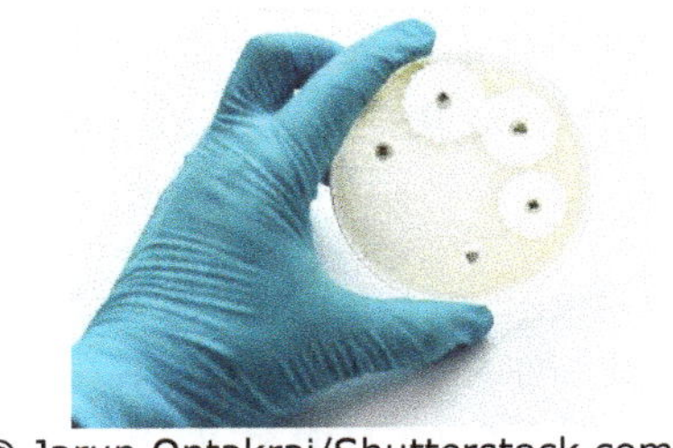
© Jarun Ontakrai/Shutterstock.com

It is important that we use **aseptic techniques** which are methods that reduce the possibility of contaminating our bacterial samples. Whatever we use to touch the microbe, such as a cotton-tipped applicator in today's lab, should be sterile and after use, should be disposed of appropriately.

Today, you will 1) determine if there are microbes on your hands, 2) if there are microbes on various surfaces on our campus, 3) determine which antibiotic would control a microbe such as *E. coli* and 4) observe a gram staining procedure.

Procedure 1

1. Each person should obtain one TSA petri plate and press the fingers of one hand gently to the surface of the plate.
2. Cover the plate and using a marker to label the bottom of the plate with your initials and the words "Hand sample".
3. Place the plate in an incubator upside down.
4. NEXT LAB SESSION- Open the plate and observe any bacterial **colonies** (areas of bacterial growth).

Procedure 2

1. Obtain one TSA plate and one SDA plate **for your group**.

2. Sample various surfaces (door knob, water fountain handle, toilet handle, computer mouse, pens etc.) by pressing the surface of the petri plate media to that surface. Sample one surface per plate.
3. Cover the plate and using a marker, place your initials on the bottom of the plate and indicate what surface was sampled.
4. Place the plate in the incubator.
5. NEXT LAB SESSION- Open the plate and observe any colonies.

Procedure 3

1. Obtain one MH plate and a sterile cotton-tipped applicator.
2. Carefully, dip the applicator into the tube that you obtained from your instructor which contains a bacterial sample (*E. coli*).
3. Using the applicator, spread the sample over the surface of the plate making sure to cover **the entire surface**.
4. Replace the cotton-tipped applicator into its paper sleeve cover and dispose of it into the red plastic bio-hazard bag.
5. Cover the plate for 5 minutes.
6. Obtain 3 antibiotic dispensers, a beaker of alcohol and forceps.
7. Dip the forceps into the alcohol and shake the forceps.
8. Use the forceps to obtain a paper disk from the antibiotic dispenser and place it on the surface of the petri plate. Tap the disk firmly onto the surface of the media.
9. Repeat steps #8 and 9 with the other 2 antibiotic disks.
10. When finished, dip the forceps into the alcohol and shake the forceps.
11. Label the bottom of the plate with your initials and the microbe " *E. coli*".
13. Place the plate in an incubator.
14. Wash your hands thoroughly.
15. NEXT LAB SESSION: Open the plate to determine if there are zones of inhibition surrounding the disks.

Which antibiotic(s) would be effective against *E. coli*?

____

Procedure 4 – Considering the Steps in Gram Staining

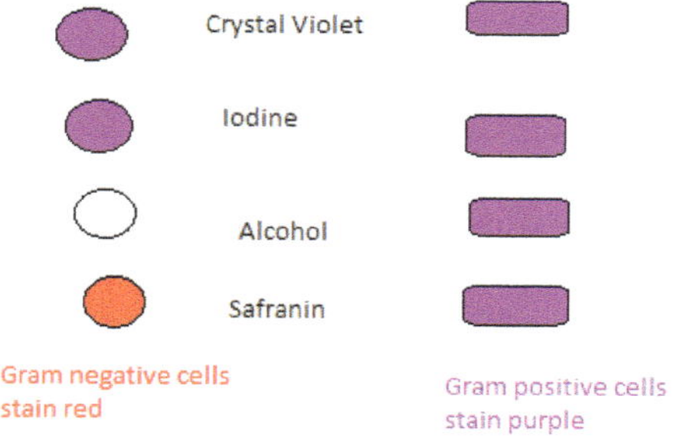

Source: Werner Williams

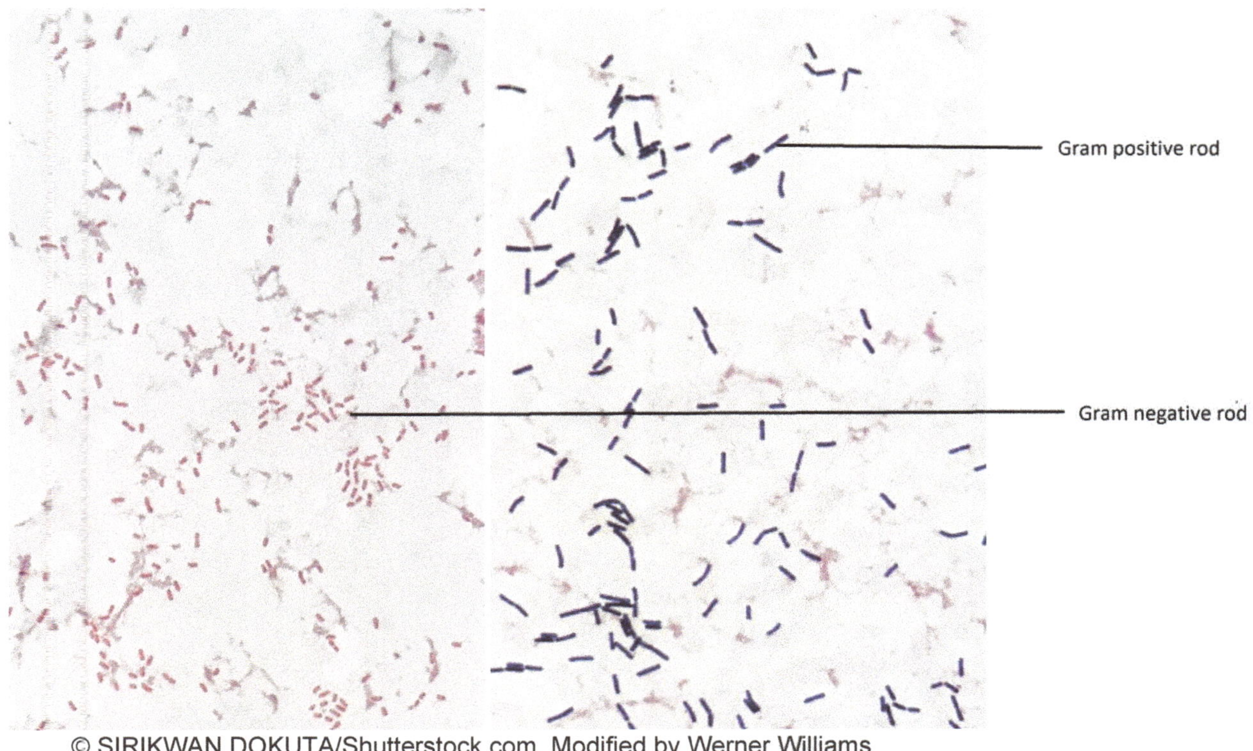

© SIRIKWAN DOKUTA/Shutterstock.com. Modified by Werner Williams

1. *E. coli* is red after gram staining. Does that make it Gram + or Gram -? __________
2. *Staphylococcus* is purple after gram staining. Does that make it Gram + or Gram -? _______________
3. Would penicillin tend to be more effective against *E. coli* or *Staphylococcus*? _____________________
4. Which microbe *E. coli* or *Staphylococcus* would have a thicker peptidoglycan? ________________________________

Lab 12: Dissection of the Fetal Pig (Parts 1 & 2)

Introduction

Mammals are different from other animals with backbones (vertebrates) in at least two ways. Firstly, they have hair, even though your fetal pig probably does not have much because it is not as yet fully developed. And secondly, all female mammals have mammary glands for nourishing their young.

We are using fetal pigs in our lab because of their availability, manageable size, low cost and their anatomy is very similar to ours. The fetuses we are using come from mature pigs that when slaughtered for meat are discovered to be pregnant. The young develop in their mothers for about 16 weeks and the ones we are using should have fully developed organs.

External Anatomy

Like most mammals, the body of a pig is divided into three sections: **the head, trunk and tail.** Figure 1. The trunk is further divided into the **thorax** and **abdomen**. The thorax contains the heart & lungs while the abdomen houses the digestive, excretory and reproductive organs. Both males and females have nipples in the abdominal region near the umbilical cord. In females, they develop into the mammary glands which will produce milk to feed their young, while in males, they have no known function.

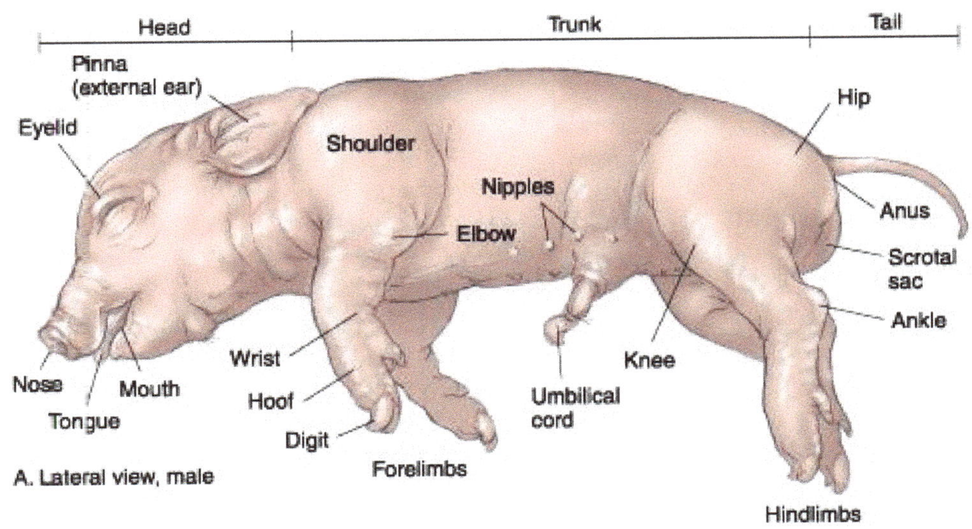

© Kendall Hunt Publishing Company

Figure 1 External Features of the Pig

Dorsal refers to the section of the pig that is toward the back or top of the body and **ventral** refers to the section toward the belly. **Anterior** refers to the section toward the head and **posterior** refers to the section toward the tail. Figure 2. When we refer to the right or left, we are referring to the pig's right or left.

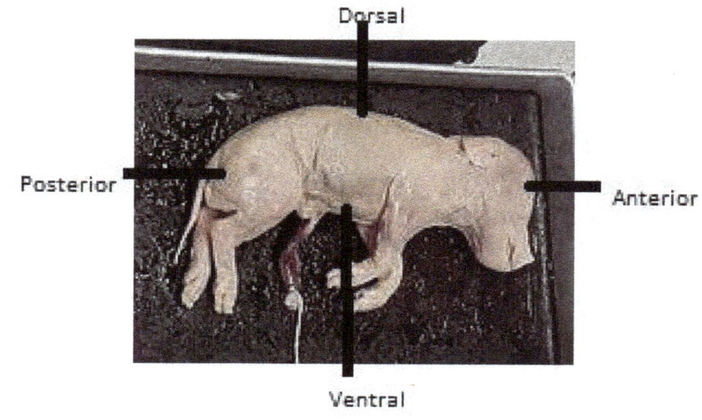

Source: Werner Williams

Figure 2 Anatomical Positions of the Pig

If your pig is an old enough male, it should have a **scrotum** containing the **testes**, located near the anus. It also has a **urogenital opening** near the umbilical cord. The term urogenital refers to the double function of this opening, because it is through this opening in males that urine and sperm are released. Females have a urogenital opening close to the anus and a short finger-like projection, the **genital papilla**. Figure 3. What is the sex of your pig? ____________________

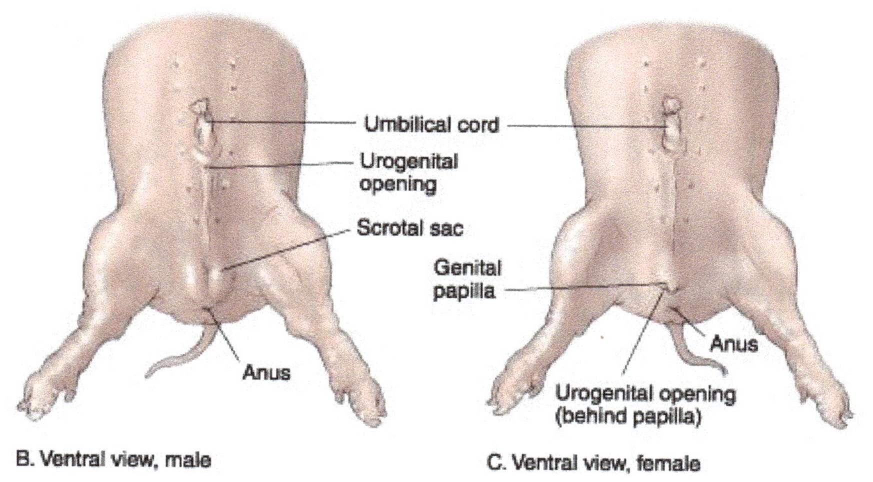

© Kendall Hunt Publishing Company

Figure 3 View of Male and Female Pigs

The Oral Cavity

Place the scissors into the mouth of the pig and cut through the cheeks toward the ear, following the cuts shown in Figure 4. **Be careful where you put your fingers because the teeth are very sharp.** You will be cutting through tissue and bone. Alternate between the right and left sides until you can open the mouth very wide. Cut both sides of the mouth until the epiglottis "pops out" in the center of the throat area.

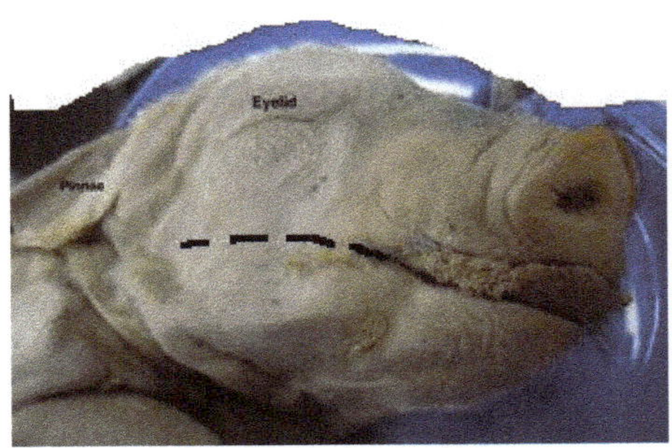

Source: Werner Williams

Figure 4

You should see **teeth** protruding from the roof of the mouth, which is composed of a bony **hard palate**. The **soft palate** continues down from the hard palate and is fleshier than the hard palate. The opening to the **esophagus** should be visible. The opening into the larynx called the **glottis** is protected by a flap of cartilage called the **epiglottis**, so that when the pig swallows, food does not travel down the glottis to enter the lungs. On the lower jaw is found the **tongue**. Figures 5**.**

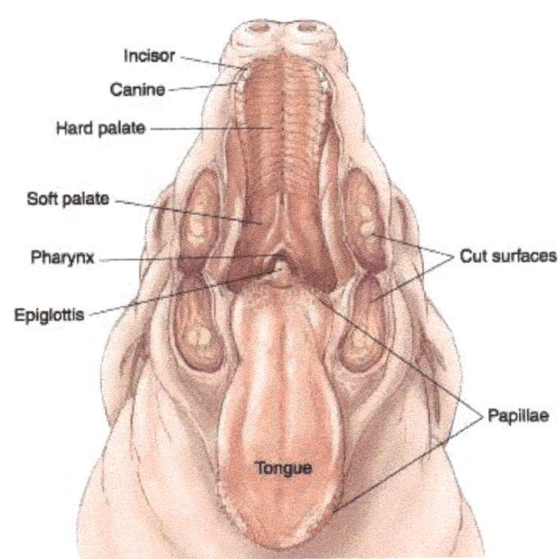

© Kendall Hunt Publishing Company

Figure 5 Oral cavity

The Body Cavity

Using a scalpel and scissors, make the cuts seen in Figure 6, to open the body cavity. Use the scissors to cut through the sternum and rib cage.

A thin muscle called the **diaphragm** separates the upper **thoracic cavity** from the lower **abdominal cavity**. In the thoracic cavity, a thin **pericardial membrane** surrounds the heart and a **coronary artery** may be seen on the surface of the heart. You should be able to locate the upper **atria** and lower **ventricles** of the heart. The **lungs** may be seen on either side of the heart and are composed of many lobes. Figure 7.

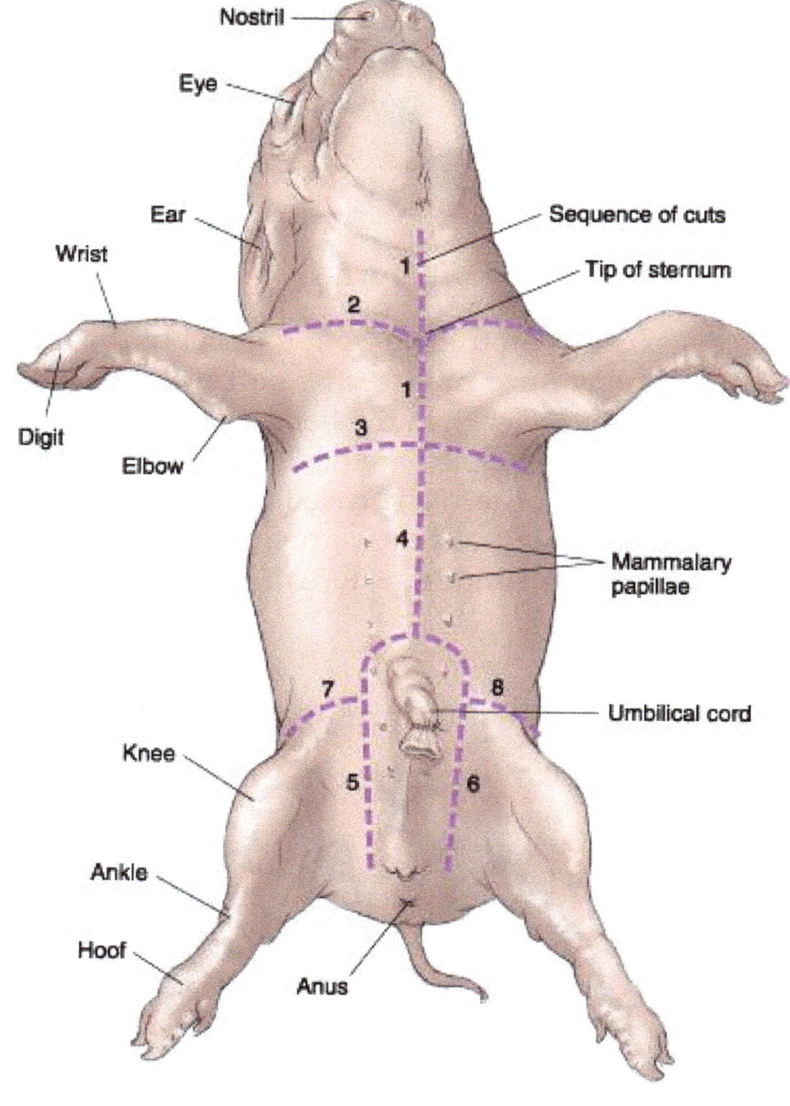

© Kendall Hunt Publishing Company

Figure 6 Body Cavity Incisions

Food that is swallowed travels down the **esophagus** and into the **stomach** which lies on the left side of the pig, underneath the large reddish- brown **liver**. The liver which is the largest organ in the abdomen produces bile which helps to digests fats. The bile is stored in a small pear-shaped **gallbladder** which is found on the underside of the right lobe of the liver. The gallbladder releases bile into the duodenum, the first part of the small intestines.

The stomach is a sac that stores food and begins the digestion of proteins. It empties its contents into the **duodenum**.

The long, dark red organ that is found on the left side of the stomach is the **spleen**. It removes worn-out and damaged red blood cells from our bloodstream.

The **pancreas**, a whitish-yellow, granular organ is located between the stomach and the small intestines. It releases digestive enzymes into the duodenum helping to digest the food, and it also produces the hormone insulin.

The **small intestines** is where most of the digestion of food takes place, breaking down proteins, carbohydrates, fats and nucleic acids. The digested food is then sent into the bloodstream to supply the cells with nutrients.

The **large intestines** is on the left side and is where most of the absorption of water occurs. The lower section, the **rectum** stores the undigested material until eliminated through the anus. Figures 7 and 8.

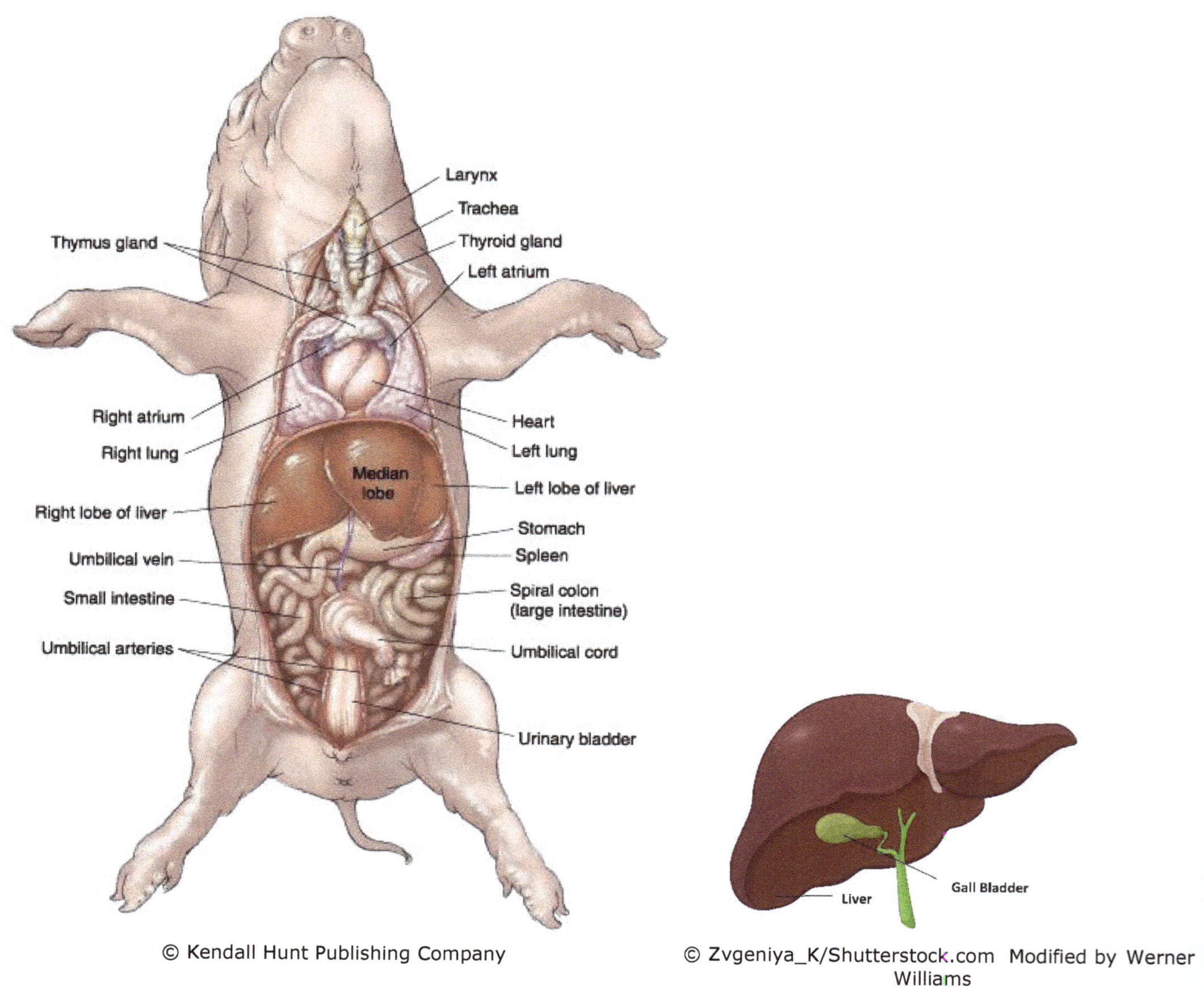

© Kendall Hunt Publishing Company

© Zvgeniya_K/Shutterstock.com Modified by Werner Williams

Figures 7 & 8 Organs of the Thoracic and Abdominal Cavities

Questions

1. The three main parts of a pig are the ____________________________________

   ____________________________________.

2. The trunk is divided into ____________________________________
3. What are two characteristics of mammals? ____________________________________

   ____

4. The word that is used to refer to the section of the pig that is toward the head is________________________________
5. The word that is used to refer to the section of the pig that is toward the tail is ________________________________
6. The word that is used to refer to the section of the pig that is toward the belly is ________________________________
7. The word that is used to refer to the section of the pig that is toward the back or top of the body is________________________________
8. The urogenital openings in male pigs is found ____________________________

____

9. A scrotum is found in______________________(male, female) pigs.
10. A genital papilla is found in__________________(male, female) pigs.
11. The flap called the___________________________closes the glottis so that food does not enter and travel down into the________________________.
12. The bony roof of the mouth is called the___________________________.
13. The muscle that separates the abdominal cavity from the thoracic cavity is the_____________________________.
14. The membrane that surrounds the heart is called the ____________________
15. The lower chambers of the heart are called the ___________________________

and the upper chambers are called the_________________________.

16. The organ that mainly functions to store food is the ______________________
17. The nutrient that begins being digested in the stomach is _________________
18. Bile is made by the___________________________and is stored in the

____

19. The nutrient that bile breaks down is __________________________
20. Insulin and digestive enzymes are both produced by the _________________
21. Most of the digestion of food takes place in the _______________________
22. The first part of the small intestines is called the _________________________
23. Most of the water in the digestive system is absorbed by the

____

24. The tube that takes food from the throat to the stomach is called the

____

25. Worn out blood cells are stored in the __________________________ which is an organ found to the left of the stomach.

Lab 13: Dissection of the Fetal Pig (Part 3)

The reproductive system is primarily responsible for producing the egg and sperm (gametes). The excretory system eliminates the metabolic wastes that are generated in the body and helps maintain a constant internal environment of water and other chemicals. This lab is primarily focused on the organs of the reproductive and excretory systems.

Procedure

1. Using scissors, remove the intestines from the body cavity.
2. You instructor will provide you with additional information on where to make the next incisions (cuts).

Excretory System

The **kidneys** are large bean-shaped organs on either side of the spine along the dorsal surface of the abdominal cavity. They filter blood, removing metabolic waste products while conserving water, salts and other chemicals like glucose. They also maintain the blood pH. In humans, the kidneys filter from 290 to 528 gallons of blood each day but only about 0.4 gallons of urine is produced. The other 99.9% is reabsorbed back into the bloodstream. The urine is concentrated in the kidneys and flows down each **ureter**, which takes the urine to the **urinary bladder** for storage. The urine then exits the bladder through the **urethra** to the outside.
Figures 1 and 2.

Male Reproductive System

In mature pigs, the **testes** are small bean-shaped structures that produce sperm. Since sperm production is very sensitive to temperature, the testes in most mature mammals are found in the **scrotum** where the temperature is a few degrees cooler than in the body cavity. The testes develop in the abdomen and then migrate into the scrotum just before birth. Sperm are stored in the **epididymis.** In an ejaculation, the sperm travels from the epididymis, through the vas deferens and **urethra**, which is a tube in the penis, to the outside. A number of different glands (prostate, seminal vesicle, and bulbourethral gland) produce a fluid called semen which transport the sperm through the urethra and into the female reproductive tract. Figure 1.

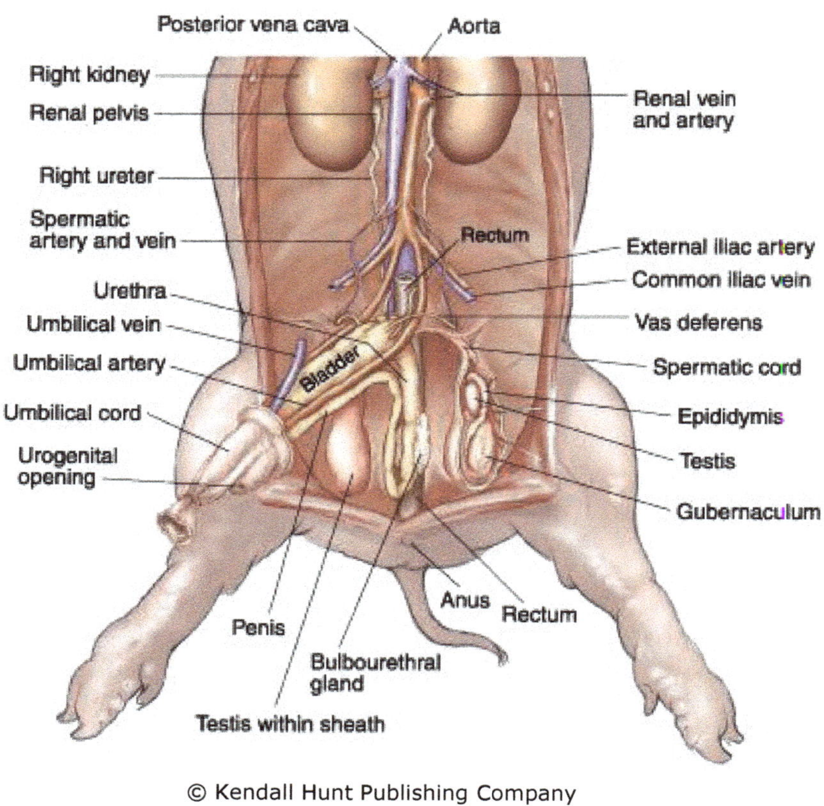

© Kendall Hunt Publishing Company

Figure 1 Male Reproductive and Excretory Organs

Female Reproductive System

The paired **ovaries** are found in the abdominal cavity, below the kidneys. They produce the **eggs** and each ovary is connected to a tiny coiled **oviduct** which receives the egg during ovulation. Fertilization usually occurs in the oviducts and the embryo implants in the upper part of the **uterus** in pigs called the uterine horns. The lower part is the body of the uterus. In human females, the embryo implants in the body of the uterus. The lowest part of the uterus is the **cervix** and is connected to the **vagina** which extends downward. The vagina, which receives sperm from the penis joins with the urethra, and the two open into a common chamber called the urogenital sinus (vaginal vestibule) which opens to the outside of the body through the urogenital opening. The urethra and vagina in human females have separate openings to the outside of the body close to each other. In the fetal pig, there is a **genital papilla**, which is a finger- like projection on the lower surface of the abdominal cavity close to the urogenital opening.
Figures 2.

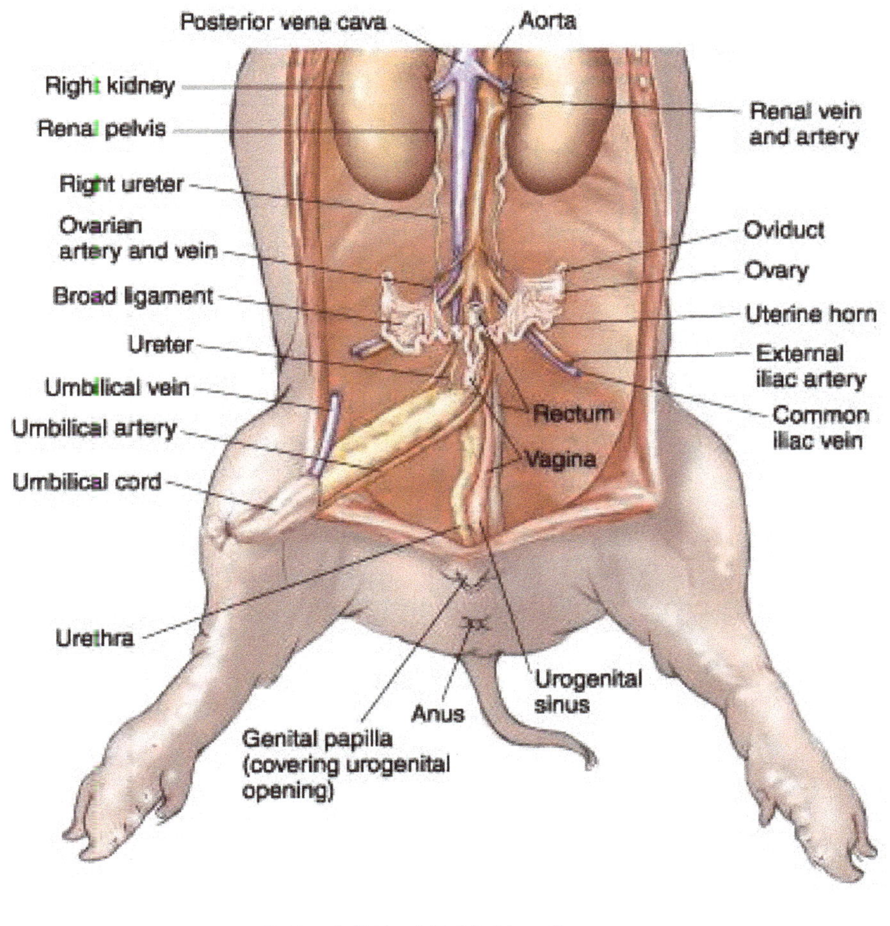

© Kendall Hunt Publishing Company

Figure 2 Female Reproductive and Excretory Organs

Questions

1. The organs which are found in the abdomen on either side of the spine that filter the blood to remove wastes and are called ____________________________
2. The tubes that directly drain the kidneys are called the ___________________
3. The tube that directly drains the urinary bladder is called the _____________
4. The sac that stores urine is called the ____________________________________
5. The name of the organs that produce sperm. __________________________
6. The tube where sperm is stored. _________________________________
7. Are the testes found inside or outside the adult male body? Why?
____
____

8. Name some of the glands that produce semen. ____________________________

____

9. Name the sac where the testes are found. ______________________________
10. Name the organs in females that produce eggs. __________________________
11. The name of the lowest part of the uterus. _____________________________
12. The name of the tube which connects the ovary with the uterus.

____

13. Where exactly does fertilization usually occur? ___________________________
14. The developing embryo implants in human females in the _________________
15. The tube in females which receives sperm from the penis is the

____

16. This is a finger-like projection in female pigs close to the urogenital opening.

____

www.ingramcontent.com/pod-product-compliance
Ingram Content Group UK Ltd.
Pitfield, Milton Keynes, MK11 3LW, UK
UKHW051326070726
13610UKWH00014B/95

9 781524 961534